AF147433

Climate Change Discourse in China

Sidan Wang

Climate Change Discourse in China

Sidan Wang
China Foreign Affairs University
Beijing, China

ISBN 978-981-16-6753-4 ISBN 978-981-16-6754-1 (eBook)
https://doi.org/10.1007/978-981-16-6754-1

Cover illustration: © Melisa Hasan

This Palgrave Macmillan imprint is published by the registered company Springer Nature Singapore Pte Ltd.
The registered company address is: 152 Beach Road, #21-01/04 Gateway East, Singapore 189721, Singapore

Preface

Establishing this book has been across my Ph.D. study and my current work. It has witnessed many critical moments in global climate governance and China's low carbon actions. With the development of global climate politics particularly since the 2015 Paris climate conference, the information of this book has been updated over time. Also, I had gained great opportunities to learn a lot from various academic scholars. For example, a paper around the rise of China in global climate governance (see Chapter 2) was presented at the Hallsworth Conference on China and the Changing Global Order in Manchester in March 2017. And, the research around framing climate resilience in China (see Chapter 4) was presented at Penryn Campus, University of Exeter, in July 2019, when I made a trip to attend my Ph.D. graduation ceremony. My Ph.D. colleagues and great friends had made their comments on my research from different academic subjects including politics, bioscience and energy.

This work can be seen as a significant milestone in my life as it is the first book I publish. Also, it has witnessed my Ph.D. study and the beginning of my career life. In this sense, I must give big thanks to the Environmental and Sustainability Institute, University of Exeter, and the Institute of International Relations, China Foreign Affairs University. My thanks also to Jacob Dreyer, a very nice and professional editor, and

Professor Clare Saunders, my Ph.D. supervisor who guides me to enter the field of climate politics. A sincere thank is to my father, Mr. Qing Wang, who gives me a strong support for my research.

Beijing, China Sidan Wang

Contents

About the Author

Dr. Sidan Wang is a Lecturer at the Institute of International Relations, China Foreign Affairs University. He received his Ph.D. in Politics at the University of Exeter in 2018. His research and teaching courses are around non-traditional security studies, environmental and energy diplomacy, and climate diplomacy.

List of Figures

List of Tables

CHAPTER 1

Introduction

> We should create a future of win-win cooperation, with each country making contribution to the best of its ability. For global issues like climate change, a take-more-give-less approach based on expediency is in nobody's interest. The Paris Conference should reject the narrow-minded mentality of "zero sum game" and call on all countries, the developed countries in particular, to assume more shared responsibilities for win-win outcomes.
>
> —President Xi's speech at opening ceremony of the 2015 Paris climate summit

Urgency and Challenges of Climate Change

Climate change has been posing a threat to the human society and our planet. According to the fifth report of the Intergovernmental Panel on Climate Change (IPCC), human activities are identified as a clear influence over the climate system, the emissions of greenhouse gases are recorded in the highest level and climate change has been observed to have obvious impacts over the ecological systems (IPCC, 2014). The sixth IPCC report on the physical science basis enhanced the confirmation of

S. Wang, *Climate Change Discourse in China*,
https://doi.org/10.1007/978-981-16-6754-1_1

climate challenges and described the human influence over climate system as an unequivocal reality. Many changes caused by the emissions are irreversible within next few centuries and millennia (IPCC, 2021). The clear and sounding scientific findings have implications for a stronger global action on addressing climate change issues. However, climate disasters have been widely affecting human communities and ecological systems across national borders. In 2019, the rainforest fire in Brazil turned global attention to the deforestation in Amazon. The Amazon is recognised as the lungs of the planet and thus it plays a key role in stabilising the carbon level in the atmosphere. And, Australia had suffered from the unprecedented forest fires. The fires posed a direct threat to the wild animals and local communities, and the haze had changed the routine for life and work in the urban regions such as Sydney. In 2021, the hottest summer strongly hit the Northern Hemisphere. From the Western Europe to the Northern America, the communities and systems had been 'fried' by the hot summer. In addition to the high temperature, the severe flooding causes disastrous impacts in Germany, Belgium and other Western European countries. It is really important to note that these extreme weather events and natural disasters had occurred in the developed countries and major economies. Clearly, climate change affects the common interests of the global communities.

China is a victim of negative effects of climate change issues. In January 2008, the Southern China had been hit by the heaviest snow within the five decades. The extreme weather events had strongly affected the national transport system and disrupted the food and energy supplies. They came when the Chinese people were preparing for celebrating the Lunar New Year which is the most important traditional festival in China (ChinaDaily, 2008). This case has raised the urgency of climate change issues particularly in policy decision-making in China. Also, the interactions between climate change, food supplies and energy security have been recognised. The southern regions of China require a coal supply from northern provinces to support the local electricity system. The heavy snow disrupted the railway system affecting the stable supply of coal. Highways were closed and roads were blocked affecting the food supplies. The year 2008 also witnessed the establishing of the links between climate and security. *The 2008 White Paper on Climate Change* recognised climate change as a threat towards China. *The White Paper of China's National Defence in 2008* raised the term 'climate change issues' (Bo, 2016: 103).

In contrast to the scientific findings of climate change, global climate governance has been developing and progressing in a disappointing way. Institutionally, the international society has established the United Nations Framework Convention on Climate Change (UNFCCC), the Kyoto Protocol and the Copenhagen Accord. However, the US, the largest carbon emitter at that moment, refused to ratify the Kyoto Protocol and failed to undertake the leadership in global climate governance. The gap between developed and developing countries had been widening in terms of climate mitigation, adaptation, financial and technological supports.

Since 2012, President Xi Jinping has raised the construction of ecological civilisation as the top priority on political agenda in China. Under close cooperation between President Xi and President Obama, China and the US made a great contribution to the establishment of the Paris Climate Change Agreement in 2015. In 2017, while President Trump required the US to withdraw from the Paris Agreement, China and other major economies and emitters insisted on supporting and continuing the global climate action. In 2020, President Xi announced the objectives of achieving the carbon peak before 2030 and the 2060 carbon neutrality (Xinhuanet, 2021). This demonstrates historic moments of climate governance of China as it undertakes the great responsibility for addressing climate change issues.

The Role of China in Global Climate Change Governance

China is a victim of climate change confronting the common challenges across national borders. In terms of climate risks, China was ranked as 31st in 2017 and 37th for 1998–2017 (Eckstein, 2018: 28–35). While China has been a developing country for a long term and thus it does not have historical emissions and responsibilities, it has to confront the climate impacts. This is an explicit reason why China supports the global climate action and international institutional cooperation. At a national level, climate change adaptation has been implemented in the sectors of agriculture, water resources, forestry, ecosystems, coastal regions, public health, disaster management and risk control. Drought, flood and extreme weather events caused by climate change were identified as potential risks (MEE, 2019: 12–16). While China has witnessed a rapid economic growth and progress in social development, it remains behind the developed countries in terms of financial and technological capabilities of mitigating climate change.

While China does not have substantial historical emissions of greenhouse gases, it is the largest carbon emitter at the moment, and it witnesses an increase in per capita emissions. The US, followed by the UK and Germany, plays a leading role in carbon emissions between 1850 and 1960. In contrast to this, China had recorded a very low level of emissions during the period. For example, in 1949, when the People's Republic of China was founded, its level of carbon emissions was recorded as 83.6 Mt while the US and the UK reported 2.83Gt and 501Mt, respectively (WRI, 2021). However, China's carbon emissions have been growing substantially since its success in economic and social developments. China's per capita carbon emissions were recorded to be much higher than those of other major developing countries particularly Brazil and India. China reached the level of the EU while it remains far behind the US and Canada (WEF, 2019).

China has witnessed its rapid economic growth. On the one hand, behind the robust economic performance, environmental costs and energy supplies have been emerging onto China's political agenda. The economic growth has increased energy consumption and required diversification of energy sources. And, industrial emissions pose a threat to environmental protection. Therefore, China started the targets of energy saving and pollution reduction in the early 2000s (Tsang & Kolk, 2010). On the other hand, China's economic development enhances its financial and technological capabilities of addressing climate change issues. In July 2021, China opened the transactions in the national carbon trading market having implications for a further development of green finance. Also, China has become the largest actor to have investments in renewable energy (IRENA, 2021). In this sense, China has been required to undertake global leadership of and responsibilities for addressing climate issues.

Indeed, China has played a proactive role in global climate governance. In 2007, China issued *the National Climate Change Programme* outlining the climate objectives. Since 2012, President Xi has raised ecological civilisation as one of core political concepts. China signed and ratified the 2015 Paris Climate Agreement and achieved its 2020 climate objectives and commitments. President Xi had attended the UN Summit on Biodiversity in September 2020, the Climate Ambition Summit in December 2020 and the Leaders Summit on Climate in April 2021.

China has promised to strengthen its 2030 climate objectives and achieve the 2060 carbon neutrality. Also, it raised a conceptual framework for building a community for man and nature as an international ecological solution (ChinaDaily, 2021).

South-South cooperation is a fundamental framework to guide China to implement its supports and assistances to developing countries. The China-Africa Environmental Cooperation Centre was scheduled to be established in 2018 as a political outcome of the Beijing Summit of the Forum on China-Africa Cooperation. The United Nations Environment Programme (UNEP) is in charge of supporting the operation of the Center and ensuring the technology support and capacity building under the framework of South-South cooperation. One of objectives of the Center is mobilising the environmental funds (UNEP, 2018). Another example of South-South cooperation is China's engagement in climate projects in Pacific island countries (UNDP, 2017). A distinctive feature of China's contribution to climate actions in developing countries is South-South cooperation. This shows a fundamental difference to the historical emissions of the developed countries and their responsibilities for providing financial and technological supports to developing countries.

A Discursive Approach to Climate Governance

There are various approaches to explaining the climate change politics and governance of China. Existing studies have employed material, structural, institutional, organisational and cultural approaches. The material approach refers to the objective impacts of climate change and the economic costs of a low carbon transition. While the climate risks have been recognised explicitly in China, a cost–benefit assessment of the low carbon action is a key element in the climate governance. While economic tools particularly a national carbon market demonstrated China's recognition of benefits from low carbon development, it is quite important to find out how ideas for low carbon development have been changing and affecting climate governance (Lo, 2015). The structural approach focuses on how international relations and structure determine climate policy options of China. Global climate politics is very different to traditional military conflicts and security concerns. While major powers have determined global military security and international financial systems, they do not necessarily represent their carbon emissions in global climate leadership (Case et al., 2015; Karlsson et al., 2011). The US is a striking example of failing to undertake its responsibility for leading the low carbon action. In this sense, the (re)distribution of global political power has a weak role in explaining the changes in climate governance of China. The institutional approach looks at the evolution of formal governmental structures of addressing climate change issues in China (Tsang & Kolk, 2010). The leading agency of making climate policies had been transferred from the National Development and Reform Commission (NDRC) to the Ministry of Ecology and Environment (MEE). The institutional construction is on the basis of policy ideas and political discourses. The process of translating various ideas into actions remains unclear. The organisational approach has been employed by studies on public administration and policies focusing on specific climate policies and measures and stakeholders involved (Bernauer et al., 2016; Schreurs, 2017). This reflects an important dimension for discovering the competing policies and stakeholders identified in the climate governance. But, the extent to which competing and various ideas have affected climate policies requires a further research. The cultural approach combines policies and institutions as a perspective for understanding low carbon politics (Toke, 2018). This raises the important role of cultural concepts and public behaviour in understanding climate politics.

A discursive approach can be used to enhance and support, rather than undermining, the existing approaches to understanding climate governance of China. Theoretically, the discursive approach has been widely used in studies on environmental politics (Hajer & Wytske, 2013). Discourse has a power to enact agents to affect environmental policies and institutions. Different stakeholders might have various ideas representing their interests, beliefs and political views. Discourse plays a key role in raising different stakeholders to emerge in climate positions. For example, oil giants have embraced climate discourses and low carbon industrial strategies. Also, discourse entails a range of different concepts and ideas. It is very important to observe how the competing discourses have been finally translated into policies and politics (Dryzek, 2013). It is interesting to explore why some ideas and discourses have not yet been dominant in climate governance in China. For example, as Chapter 4 implies, while climate resilience has been adopted clearly in climate policies particularly in European countries, it remains very inconsistent in its interpretation of core concepts and definitions in the context of China. External events and critical policy moments have been discursively interacting with the climate ideas and concepts. The discourses have to be observed and analysed in different contexts (van Dijk, 1997). For example, climate justice and responsibilities for the low carbon action have been (re)interpreted in different ways.

Practically, the discursive approach has an inherent and critical value of studying climate governance of China. First, the approach provides a dynamic perspective to observe the climate governance. It is difficult to define the motive for China's climate policies in one way. With the changing of discourses, the motive and interests have been interpreted in different ways and have been evolving over time. Second, various ideas have been framed to guide climate governance and policies. An issue can be framed as problems, causes, moral responsibilities and practical solutions (Entman, 1993). A dominant frame can be identified as an important contribution to the shaping of climate policies. For example, framing climate justice has implications for judging the stakeholders to be responsible for taking actions. Third, it is very important to focus on the discourses beyond governmental policies. While the Chinese administrative system plays a key role in governing climate issues, various actors have been widely involved in low carbon actions. Financial and business agencies have incentives to raise green finance projects, and environmental non-governmental organisations (NGOs) have participated in constructing green concepts and influencing public understandings of environmental issues. This research summarises and defines the climate discourses from the governmental statements and policies, and it also observes how various actors have been speaking for climate change issues in the context of China.

How to Understand Climate Governance of China?

While the discursive approach provides a very powerful framework to analyse a wide range of climate-related topics and fields, this book focuses on the widely focused and emerging themes around the climate governance. First, climate leadership has been employed as a theoretical framework to map the complex climate governance of China. It is very important to note that a discursive approach to climate leadership shows an innovative perspective for understanding the climate politics in China. Not surprisingly, the decision-makers at the central level play a key role in the climate governance. However, the power of leading climate policies has been redistributing across different functional governmental bodies.

The MEE undertakes the leading role in climate policy-making from the NDRC signifying a milestone in China's low carbon and climate actions. This demonstrates a discursive shift from economic priority to ecological conservation. Also, the redistribution of climate leadership has been observed across vertical and territorial administrative systems in China. This is not to say that the central leadership has been undermined in China's low carbon actions. Instead, the redistributed and shared leadership across different administrative levels offers the local actors to have their voices and interests in climate governance. This is helpful to optimise and smooth the implementation of climate policies. In addition to the governmental bodies, various actors are observed to have their ideas and voices in the climate discourses and governance. On the one hand, the non-state actors can be followers to the climate leadership at the central governmental level of China. Particularly, they make a contribution to communicating and supporting the key concepts of ecological civilisation, carbon peak and carbon neutrality. On the other hand, they have shared green ideas with other actors and governmental agencies. The term 'low carbon development' has been emerging across business and environmental agencies and finally been adopted in China's governmental discourses. The shared and redistributed climate leadership has enhanced, rather than undermining, China's climate actions. The dynamic climate leadership of China has not only been translated in the climate policies but also it has reflected China's participation in global climate governance.

Second, China's role in global climate politics is also determined by China's policies on diplomacy and international cooperation. An internal contribution to the rising role of China in global climate governance can be demonstrated by domestic climate actions including establishing targets of controlling energy intensity, carbon intensity and carbon emissions. Also, it is important to observe the evolution of China's attitudes towards international cooperation. The 2008 global financial crisis has a profound effect on global political and economic structures. While the developed countries and major powers have failed to address global challenges including financial crisis, terrorism, climate change and public health issues, the emerging economies have been recognised as an important role in enhancing global governance. The G20 has become an important multilateral platform to address global challenges. It challenges the traditional Western mechanisms of managing global affairs such as the G7. Since 2008, China has started to have a more proactive role in participating in international affairs. As Chapter 2 outlines, China has clarified the fundamental policies of developing diplomatic relations with major countries, neighbouring countries and other developing countries. This diplomatic position can be more or less reflected in China's role in climate governance. Under the close cooperation between President Xi and President Obama, China and the US made a great contribution to the 2015 Paris Climate Conference. On the other hand, China has proactively been working with Brazil, India and South Africa to defend their common positions in climate negotiations. While China has been showing its active attitudes to global climate governance, it insists on the fundamental principles of addressing climate issues under the framework of the United Nations.

Third, the principle of common but differentiated responsibilities is recognised as a fundamental position in having a global collective action on addressing climate issues. This is also a political basis for international climate cooperation. The developed countries have historical responsibilities for addressing climate issues and they have national capabilities to provide financial and technological supports to developing countries and vulnerable communities. However, various voices have been emerging around discussions on the principle in China. It is very interesting to observe the discursive evolution of climate justice and its responsibilities. While the fundamental principle has been rooted in China's climate positions, the voices for the responsibilities of China and major emitters have been growing. China's economic growth and carbon emissions have been cited by different stakeholders to construct China as a major emitter and an important contributor to climate change. While climate actions and justice have received attention widely, climate resilience has remained less important than other topics on climate discourses in China.

Fourth, practically and technically, climate resilience plays an important role in the governance, while it has not yet triggered citizens, various stakeholders and various governmental agencies to recognise its importance and urgency. On the one hand, mitigation is a significant step towards conserving the planet. On the other hand, adaptation has a profound effect on climate justice and human survival. Discursive construction of climate resilience determines how agencies are going to take actions and respond to the climate crisis. This book has a very important contribution to studies on discourses on climate resilience. As Chapter 4 finds, the Chinese characters 'resilience' have been constructed in a different way to the term in English. The different understandings and interpretations might have different implications for governmental and societal responses to climate disasters.

Structure of the Book

Chapter 2 focuses on the climate leadership in China and employs a state transformation approach to observing the various agencies in climate governance. The approach offers a dynamic perspective for observing the redistributing and sharing of the climate leadership across different levels. The chapter discusses the administrative systems for climate governance and also maps the distribution of various stakeholders around the climate leadership. Following the mapping of various agencies related to climate governance, it analyses the role of the Ministry of Ecology and Environment in China's climate governance and leadership.

Chapter 3 focuses on China's rise in global climate governance. It reviews the evolution of China's engagement in global climate politics. By analysing the governmental statements, it compares the China's positions on international cooperation and climate change in 2009 and in 2015, respectively. This chapter argues that the domestic political discourse on the rise of China in international cooperation makes a significant contribution to Chinese attitudes towards global climate change governance.

Chapter 4 focuses on the discursive change in the principle of common but differentiated responsibilities in China. This is not to say that the Chinese governmental position on the principle has changed. However, this research identifies various stakeholders having constructed the definitions of the principles in different ways. China has been discursively constructed to undertake the responsibility for addressing climate issues. It selects the key climate statements released by governmental agencies and identifies *China Daily* as an important source to observe the discourses across different social actors.

Chapter 5 focuses on the framing of climate resilience in newspapers in China. It observes how the climate resilience has been mediated across different types of newspapers in China. Particularly, it is very interesting to find how it has been constructed differently between the English-written *China Daily* and other Chinese-written newspapers. Also, various actors have been cited as news sources to frame climate resilience. Who are the main actors of raising the concept of climate resilience? What are the associations between triggering events and the coverage of climate resilience in China? These research questions have been addressed there.

It is important to note that climate leadership, China's role in global climate governance, the principle of common but differentiated responsibilities and the concept of climate resilience have been linked rather than being constructed separately and distinctively. While the climate leadership refers to the redistribution of domestic power, the principle is a fundamental position on international climate cooperation. The rise of China in global climate governance has been determined by domestic climate policies and strategies of international cooperation. Climate resilience refers to China's actions on confronting the negative effects of climate change at local levels. Therefore, this book employs the four key and emerging themes to explore China's climate governance across local, national and international levels in a systematic way.

References

Bernauer, T., Dong, L., McGrath, L. F., Shaymerdenova, I., & Zhang, H. (2016). Unilateral or reciprocal climate policy? Experimental evidence from China. *Politics and Governance, 4*, 152–171.

Bo, Y. (2016). Securitization and Chinese climate change policy. *Chinese Political Science Review, 1*, 94–112.

Case, P., Evans, L. S., Fabinyi, M., Cohen, P. J., Hicks, C. C., Prideaux, M., & Mills, D. J. (2015). Rethinking environmental leadership: The social construction of leaders and leadership in discourses of ecological crisis, development, and conservation. *Leadership, 11*, 396–423.

Chinadaily. (2008). Heavy snow piles on the agony. *ChinaDaily*, 28/01/2008.

Chinadaily. (2021). For man and nature: Building a community of life together. *China Daily*, 23/04/2021.

Dryzek, J. S. (2013). *The politics of the earth: Environmental discourses*. Oxford University Press.

Eckstein, D., Marie-Lena, H., Winges, M. (2018). Global climate risk index 2019 Bonn: Germanwatch.

Entman, R. M. (1993). Framing: Toward clarification of a fractured paradigm. *Journal of Communication, 43*, 51–58.

Hajer, M. & Wytske, V. (2013). Voices of vulnerability: The reconfiguration of policy discourses. In J. S. Dryzek, B. N. Richard, S. & David (Eds.), *The Oxford handbook of climate change and society*. Oxford University Press.

IPCC. (2014). Climate change 2014 synthesis report summary for policymakers. *The Fifth Assessment Report of the Intergovernmental Panel on Climate Change*. Intergovernmental Panel on Climate Change.

IPCC. (2021). Summary for policymakers. In V. Masson-Delmotte, P. Zhai, A. Pirani, S. L. Connors, C. Péan, S. Berger, N. Caud, Y. Chen, L. Goldfarb, M. I. Gomis, M. Huang, K. Leitzell, E. Lonnoy, J. B. R. Matthews, T. K. Maycock, T. Waterfield, O. Yelekçi, R. Yu and B. Zhou (Eds.), *Climate change 2021: The physical science basis*. Cambridge University Press.

IRENA. (2021). *China and IRENA boost ties as leading renewables market eyes carbon neutrality goals tweet* [Online]. International Renewable Energy Agency. https://www.irena.org/newsroom/pressreleases/2021/Jun/China-and-IRENA-Boost-Ties-as-Leading-Renewables-Market-Eyes-Net-Zero-Goals. Accessed 20 July 2021.

Karlsson, C., Parker, C., Hjerpe, M., & Linnér, B.R.-O. (2011). Looking for leaders: Perceptions of climate change leadership among climate change negotiation participants. *Global Environmental Politics, 11*, 89–107.

Lo, A. (2015). *Carbon trading in China: Environmental discourse and politics*. Palgrave Macmillan UK.

MEE. (2019). *China's policies and actions for addressing climate change*. http://english.scio.gov.cn/pressroom/2019-11/28/content_75458089_8.htm. Accessed 05 January 2021.

Schreurs, M. (2017). Multi-level climate governance in China. *Environmental Policy and Governance, 27*, 163–174.

Toke, D. (2018). *Low carbon politics: A cultural approach focusing on low carbon electricity*. Routledge.

Tsang, S., & Kolk, A. (2010). The evolution of Chinese policies and governance structures on environment, energy and climate. *Environmental Policy and Governance, 20*, 180–196.

UNDP. (2017). China's South-South Cooperation with Pacific Island Countries in the context of the 2030 Agenda for Sustainable Development, Series Report: Climate Change Adaptation. United Nations Development Programme.

UNEP. (2018). *China-Africa Environmental Cooperation Centre* [Online]. United Nations Environment Programme. https://www.unep.org/regions/africa/regional-initiatives/china-africa-environmental-cooperation-centre. Accessed 20 July 2021.

Van Dijk, T. A. (1997). *Discourse as structure and process*. SAGE.

WEF. (2019). Chart of the day: These countries have the largest carbon footprints. 02/01/2019 ed.: World Economic Forum.

WRI. (2021). The history of carbon dioxide emissions. 31/03/2021 ed.: World Resources Institute.

Xinhuanet. (2021). *China goes full throttle on green energy transition to achieve carbon neutrality* [Online]. Xinhua News Agency. http://www.xinhuanet.com/english/2021-07/12/c_1310057114.htm. Accessed 13 July 2021.

CHAPTER 2

A Leadership Approach to Understanding the Transformed Climate Governance of China

Abstract A rapid economic growth raises China as a key player of global climate change politics. China has witnessed its sharp increase in greenhouse gases emissions surpassing the US to be the largest emitter at the moment. While the leadership of global climate governance has been widely discussed by international scholars, very few studies pay attention to the sub-national leadership on the climate governance. This research looks at how the leadership has been shared, reconstructed and redistributed in the governing system of China. By using the state transformation approach, the case of China has been scrutinised with employing the elements of decentralisation, fragmentation and internationalisation. This chapter finds that the leadership on climate governance of China has been redistributed vertically, shared functionally and internationalised variously. It is argued that the transformed governance seems to have not weakened but enhanced the climate leadership in China.

Keywords Leadership · State transformation · Multi-level governance · Climate governance

S. Wang, *Climate Change Discourse in China*,
https://doi.org/10.1007/978-981-16-6754-1_2

Introduction

The literature on linking leadership to climate politics remains underdeveloped, while a call for the leadership on addressing climate change can be identified widely in political practices (de Águeda Corneloup & Mol, 2014; Hjerpe et al., 2011; Tobin, 2017). The rapid economic growth raises China as a key player of global climate change politics (Belis & Qi, 2015, p. 199). Without the active participation of China identified as the largest greenhouse gas emitter of the world, the international community particularly including major economies would fail to take the collective action on addressing climate change (Harris et al., 2013, p. 292). On the one hand, China confronts an unprecedented pressure from international society with respect to climate change issues (Zhang & Zheng, 2008, pp. 8–9). On the other hand, China faces its huge difficulty reducing emissions of greenhouse gases with its heavy dependence on the consumption of coal (Held et al., 2011, pp. 12–16) and its concern about economic development (Harris et al., 2013). Obviously, it is very important to discuss the extent to which the leadership has evolved within the climate change governance of China. This chapter employs a state transformation approach to understand the multiple climate leadership shared and redistributed by various actors in China.

The complexity of leadership on climate change governance can be demonstrated by a special issue of the journal *Environmental Politics* published in 2019 (Wurzel et al., 2019a). This was an attempt to raising the concept of leadership from the view of environmental political scientists. The special issue focused on various actors of constructing and categorising the concepts of leadership; business actors raising ecological concerns and market tools (Dupuis & Schweizer, 2019); EU member states favouring economic benefits (Jänicke & Wurzel, 2019); cities confronting economic pressures (Wurzel et al., 2019b); the EU leading its followers (Torney, 2019); small entrepreneurial actors securing their sectoral transformation (Biedenkopf et al., 2019); and electricity and conventional energy sectors having different responses to climate policies (Eikeland & Skjærseth, 2019). This leads to very practical and normative considerations of how and what kinds of climate leadership could be implemented and achieved.

The research gaps between leadership and climate change are identified and explained here. First, the climate leadership can be redistributed and shared by various stakeholders beyond the governmental agencies. While a coal power plant is a contributor to climate change, a local community far away the plant can be a victim of the negative effect. The actors of causing carbon emissions might have a structural leadership with material power while the victims raise their voices of condemning the industrial activities in the climate leadership. Second, the complexity of climate change issues leads to the various voices among different social actors. The causes, impacts and solutions of climate change issues are identified across various industrial sectors rather than a single field addressed by a traditional approach to leadership. Third, a top-down approach to addressing climate change fails to implement international climate agreements particularly the Kyoto Protocol. Thus, a shift from the top-down framework was recorded in the process of global climate governance like the Paris Agreement (Wurzel et al., 2019a). The dichotomy between top leaders and followers seems to be ambiguous. At an international level, domestic politics substantially determines the ratification of the international climate agreements. The US's withdrawal from the Paris Climate Agreement is a stark example of demonstrating the domestic factors. At national and sub-national levels, different stakeholders show various voices of addressing the climate issues. While the National Development and Reform Commission (NDRC) plays a leading role in shaping climate policies, other governmental agencies have their preferences such as the Ministry of Foreign Affairs (MoFA) focusing on national sovereignty and the Ministry of Environmental Protection (MEP) raising the importance of effectively implementing climate policies (Held et al., 2011; Tsang & Kolk, 2010). Fourth, the climate leadership has been linked to major historical carbon emitters particularly the EU and the US, while the role of China, a large developing country, seems to be ambiguous. However, a rising call for the emerging carbon emitter such as India and China emerged on the debate on climate leadership. Thus, the BASIC group, including China, Brazil, India and South Africa, was created to defend the common positions on international climate negotiations (Qi, 2011a).

The EU has been depicted as a leader in achieving global climate change actions (Afionis et al., 2012; Parker & Karlsson, 2010). Being a single unit of international relations, the EU was labelled as a leader among the large developing countries such as India and China in terms of climate actions. But, a gap between the leadership of the EU and the followership of the major emitters among developing countries emerged as divergent interests and frames of addressing climate change issues between the two sides did exist (Torney, 2015). Not only does the leadership of the EU affect the climate policies across the national governments, but it also exerts an influence across the industrial sectors. The EU emissions trade system (ETS) has been developed and learned globally, and its leading role in reducing the carbon emissions from the aviation sector has been observed within the body of the International Civil Aviation Organization (ICAO) (Lindenthal, 2014). In addition to the global leadership, a regional leading role in addressing climate issues has been examined. For example, while Japan and Australia are the developed countries in the Asia Pacific region, they have the responsibility of undertaking the leadership of offering supports to the countries substantially vulnerable to climate change (Crowley & Nakamura, 2018).

There are different types of leaders with various ideas identified in terms of climate leadership (Meijerink & Stiller, 2013). In the early stage of global climate politics, the importance of leadership was examined within the international climate negotiations. There are structural leadership having capability of mobilising resources and material power, directional leadership affecting the developments of policies (Malnes, 1995) and instrumental leadership functioning the process of climate diplomacy and negotiations (Kanie, 2003, pp. 10–11; Gupta & Grubb, 2000). For example, while the US plays a weak role in the structural leadership required for substantial financial supports, it is inclined to reflect the instrumental leadership marked by the Sino-US climate cooperation (Parker & Karlsson, 2018, pp. 525–526). In addition to these categorisations, the existing study articulates cognitive leadership redefining climate-related ideas and exemplary leadership offering policy options and examples for others (Wurzel et al., 2019a, pp. 8–11).

However, this chapter places an emphasis on a discursive perspective of understanding leadership. While various actors are placed outside the formal system of administrative leadership, they can exert an influence through discourse (Bach, 2019, p. 99). Discourse can be a power to affect the politics and link various actors within a network (Hajer, 1995). Also, an understanding of the leadership can be placed within the role of followers rather than exclusively on the position of leaders. The actor of translating leadership into practices can be described as a responsible leader. This leadership is linked to the responsibility to take action and green economic growth (Eckersley, 2016). In terms of environmental leadership, leaders are defined as those having authority of positioning, conducting and addressing the environmental issues and policies (Case et al., 2015). But this categorisation of environmental leaders explicitly undermines the agency of the so-called followers. By interrogating the relations between readers and followers, the multiple leadership can be shared and even redistributed (Crossman & Crossman, 2011, p. 485). The concept of multiple leadership breaks the stark dichotomy of the leader–follower approach and is reluctant to label 'follower' (Ford & Harding, 2018, p. 15). Rather, it emphasises the distribution of the leadership among various actors identified rather than leaders and followers labelled. A critical approach to the traditional leadership has been emerging although it has not yet been applied widely (Learmonth & Morrell, 2017).

The critical approach provides a dynamic perspective of observing the changes in the context of leadership. The relations among various actors are changing and evolving over time and thus the positions of leaders and followers vary in different spatial and temporal contexts (Van Vugt & von Rueden, 2020, p. 7). The climate governance does not raise a top-down perspective but it offers an opportunity for interconnected and dynamic relations between leaders and followers. This raises a question around the way of defining a follower and understanding the followership (Wurzel et al., 2019a, pp. 13–14). In this sense, the followership has been defined in an ambiguous way while the leadership should not be defined as an even distribution of responsibility.

In fact, the leadership among the EU, the US and China is fragmented raising a concern about global climate action rather than reassuring the situations (Karlsson et al., 2012, p. 54). While the EU was seen as a global climate leader, the US received a low level of public expectations to having its leadership (Karlsson et al., 2011). In contrast to the Obama administration elevating the leadership of the US, the Trump's team abandoned this reputation. However, it is reassuring that the climate leadership has been redistributed among the groups like the EU and the sub-national actors like California (Mazmanian et al., 2020; Palacková, 2017).

While this chapter researches the case of climate politics of China, it does not completely employ the traditional approach to leadership, e.g. the structural power. On the one hand, this chapter acknowledges that the structural leadership is an analytical tool to understand the distribution of power and policy implementation across the top-down administrative system of China. On the other hand, various actors can mobilise climate-related ideas and policy options to affect and share the leadership on climate governance in China. In this sense, it is very important to break the traditional dichotomy of leaders and followers, and instead employ a dynamic perspective to interrogate whether a transformed climate leadership has delivered the effective climate governance in China. This chapter develops a core research question: What is the nature of the leadership evolving within the climate change governance of China? How have various actors utilised climate-related ideas and discourses to enhance their agency and share and redistribute the multiple leadership?

This chapter will set a section on links between climate change, leadership and China. As the governing system of China is explained, the state transformation approach is employed as an analytical approach to understanding a dynamic climate leadership in China. Following this, the next section will focus on the main findings namely centralised, decentralised, fragmented and internationalised climate leadership in China. A section on the new era of climate governance demonstrates the establishment of the Ministry of Ecology and Environment in 2018. The section on discussion and conclusion will articulate why the multiple leadership shared and redistributed by various actors enhances, rather than undermining, the effective climate governance in China.

Climate Change, Leadership and China

As Ostrom (2012) argues, a failure to achieve a macro-leadership at the global level leads to a shift towards the leadership of sub-national actors. This chapter thus looks at how the various actors have constructed the leadership on the climate governance of China. The state transformation approach is employed as an analytical framework to understand the leadership across the vertical governmental systems, functional ministerial agencies and various non-state actors in China. The approach emphasises the decentralisation across the top-down decision-making system, the fragmentation across the ministerial agencies and the internationalisation within the actors having transboundary interests.

The states have been largely seen as a key unit of undertaking the leadership. Tobin (2017, p. 28) clarifies the responsibilities of the developed states for being leaders on driving global action on reducing carbon emissions. While small island states have a very limited political capability in international relations, they gain leadership in raising the justice and moral judgement in global climate politics (Chan, 2018; de Águeda Corneloup & Mol, 2014). With a rapid economic growth and a sharp increase in carbon emissions, China has been rising to be one of key actors in global climate change governance and has been required to undertake responsibility for substantially controlling emissions (Schreurs, 2016). The studies explicitly reveal that the global leadership of addressing climate change has been shared by different states in various ways. However, this leads to a lack of a specific analysis of and discussion on how the leadership has been achieved across the different levels at a sub-national level.

Since China ended the political-economic-social order in the 1970s, on the one hand, economic policies and activities thus started to depend on the negotiations between the administrative departments at different ranks. On the other hand, the central leadership of China remains the most important power in China. The fundamental political structure has not changed with the new era of reform. This administrative system remains robust but it is, to some degree, fragmented across different governmental levels in terms of policy-making (Hensengerth, 2015; Lampton, 1987; Lieberthal, 1992; Mai & Francesch-Huidobro, 2014). However, this system in a transforming process has witnessed a support for an effective climate leadership of China. With economic and social reforms, the state transformation approach provides an opportunity to observe how the various actors influencing the climate change policy converge on the climate leadership. The approach also looks at the interactions between local, national and international levels. As the climate change governance can be largely influenced by international factors and actors, it is important to understand the policy process beyond national level. Therefore, this research discusses the extent to which the climate leadership has been constructed and evolved across the actors within such a transformed system.

Due to the nature of China's administrative system, the climate change leadership is supposed to be secured at the central level. In this system, the leaders at the top level might exert a power to control the 'followers' namely the various actors at the sub-national levels (Ford & Harding, 2015, p. 4). However, the economic reform after the late 1970s downplays the traditional top-down system. The climate governance can be fragmented across the sub-national levels. This breaks the dichotomy between the 'leaders' and 'followers' (Learmonth & Morrell, 2016, p. 257). In this sense, it is very interesting to understand how the actors across different levels especially with their different interests and preferences construct and rearrange the climate leadership in China. It is important to note that the actors outside the environmental and climate policy-making process do not necessarily influence the Chinese governmental options. Instead, they might share and support a common discourse and affect the policy rhetoric of China.

The existing studies on the climate administration and governance of China and various actors focus on, but not limited to, the urban management (Mai & Francesch-Huidobro, 2014, p. 38), the effectiveness of environmental governance (Newig & Fritsch, 2009) and the environmental model of cities (Schreurs, 2010, pp. 96–97). Given the theory such as the Multi-Level Governance employed by the studies have been rooted and developed in the Western countries, however, this research requires a theoretical travel. Theoretical travelling refers to a concern about the applications of theory between different national and political contexts (Sabatier et al., 2005, pp. 11–19, Benson et al., 2013, p. 749). The concern questions how the theory can be travelled and extended to national contexts beyond the Western world (Sartori, 1970, pp. 1034–1036). Therefore, this chapter critically uses the state transformation approach to make sense of the leadership on climate governance of China.

Analytical Framework: State Transformation Approach to China

State Transformation Approach

The Multi-Level Governance(MLG) has dominated the theoretical frameworks employed in the studies on the political and administrative system of China for a while. However, the theories are needed to develop in accordance with the development of the national circumstance of China and be tailored to explain the climate governance.

The MLG has been widely discussed and used in European politics and international relations (Piattoni, 2010, pp. 1–2; Zürn et al., 2010, p. 1). The MLG provides an important perspective of observing and understanding the interaction between different governmental layers across local, national and international levels. It is not surprising that the MLG plays an important role in improving the understandings of European affairs and politics as the process of European unification and policy-making cannot be isolated from the complicated interactions between local, national, European and international levels.

The MLG challenges a top-down approach to policy-making recognised as the conventional notion of sovereign states (Piattoni, 2010, pp. 2–3). There are two types of systems namely general-purpose and task-specific jurisdictions (Ongaro, 2015, p. 3). The general-purpose jurisdictions can be understood as territorial systems. In the case of China, they are local, provincial and central governmental bodies. The task-specific jurisdictions are functional bodies like local and provincial environmental protection bureaus and the Ministry of Ecology and Environment in China.

Firstly, China's reform particularly in economic system provides a perspective for observing climate change governance and achieves an evolution of a top-down system. Secondly, addressing climate change involves international negotiations and local implementations. The MLG can be used to map the complicated policy systems across the local, national and international levels.

However, the MLG has two limitations. First, it fails to catch a dynamic nature of climate change governance in China. Over years, China has witnessed substantial changes in its administrative systems, institutional arrangements and governmental positions in terms of addressing climate change issues. It explicitly shows a shift from a developing country with defending its position on economic priority to a proactive actor in securing the ratification of the 2015 Paris Climate Agreement. Second, the MLG focuses on the vertical administrative system but it does not pay much attention to the competitions across various functional actors. Indeed, different ministerial agencies have various positions on and interests in climate policies and governance.

Hensengerth (2015, pp. 298–302) defines the Chinese administrative system as the central leadership and the functional and territorial fragmentation in terms of the process of policy-making. Lampton (1987) reveals that China's politics was a bargaining system rather than the command one. Lieberthal (1992) stated that the formal administrative system is a unified chain of command with central leadership and several vertical ministers which control their subordinates from central to local levels. The relationship between central government and locale is reflected in two dimensions, national uniformity and provincial autonomy (Lieberthal & Oksenberg, 1990, p. 138). The existing studies on the case of China focus on the hydropower (Hensengerth, 2014, p. 56), the Three Gorges Project in China (Heggelund, 2004) and climate change in Chinese cities (Mai & Francesch-Huidobro, 2014, pp. 37–38). This helps to raise an argument that the climate change politics of China is centralised at the top level but it can be rearranged across different administrative levels.

While the dynamic approach to the governance of China focuses on functional fragmentation, it does not emphasise the international factors identified in the climate politics. Being different to other issues, climate change has been understood as an international affair requiring a global collective action on reducing emissions. In this sense, the domestic politics of climate change in China can be affected by international contexts. Therefore, the internationalisation of the issues and governance matters.

In contrast to the MLG model, the state transformation approach offers a critical perspective of understanding a dynamic and complicated climate change leadership of China across various actors. The approach is designed to understand the rising powers on the global stage by observing the fragmentation, decentralisation and internationalisation (Hameiri & Jones, 2015, p. 2). The fragmentation of the state transformation is differed to that of the FA model containing the private sectors into the governance and system (Hameiri et al., 2019, p. 1399). The fragmentation across public and private sectors reflects the reality of climate politics of China as various actors such as academic communities, business and non-governmental organisations have been engaged in the governance. The state transformation emphasises the decentralisation of the vertical administrative system. But it looks at a wider range of actors within the system such as the state-owned companies. Zhang (2019, p. 1461) explores how the China's National Nuclear Company pursues its own international agenda undermining the importance of the primary tasks set at the higher administrative level. The internationalisation represents a distinctive feature of the state transformation approach. It refers to a process of transforming the domestic goals in the international context and having interactions with the actors at the global level (Hameiri et al., 2019, p. 1401). It has been used to explain how China is engaged in global governance in accordance with its domestic policies (Hameiri & Zeng, 2019).

This chapter argues that the leadership on climate change governance in China has been fragmented functionally, decentralised vertically and internationalised variously. The various actors including the non-state ones have different interests and positions in terms of addressing climate change issues. In the vertical system, the decentralisation indicates that the leadership has moved from the central government to the various actors at local levels. The internationalised actors have constructed the leadership combining the national interests of China and global obligations of reducing emissions.

Governing System of China

Since the late 1970s, on the one hand, economic policies and activities thus started to depend on the interactions between the administrative departments at different ranks. On the other hand, the top leadership remains the most important power in China. This governing system remains centralised but it can be decentralised, fragmented and internationalised across the various actors at different governmental levels.

Lampton (1987) reveals that China's politics was a bargaining system rather than the command one. Bargaining is generated in the process of the formulation and implementation of policy. China is often understood as a country with a command system. However, this system has changed dramatically with the economic reform. Different governmental participants have their own concerns, ideas and priorities. Therefore, in the absence of a clear consensus, decentralised policy-makers have to negotiate and bargain with each other until they reach an agreement.

Lieberthal (1992) defines the formal governing system as a unified chain of command with central leaders and several vertical ministers which control their subordinates from central to local levels. As the authors indicate, however, due to the nature of vertical systems, ministries have their own interests and priorities. The fragmentation, therefore, is determined by the fact that each ministry perceives their own missions as a priority.

There are two dimensions of fragmentation in terms of political structure. Due to equivalent bureaucratic rank, functional administrations at the same level have no authority over each other. This functional system with its strict bureaucratic rank contributes to bargaining with the purpose of reaching an agreement. In addition to the fragmented links between functional administrations, the similar relationship between central control and local autonomy also exists, to some degree, because of the budgetary reform. This allows local governments to utilise financial resources outside of central budgetary plans and to be less sensitive to central decisions (Lieberthal, 1992, p. 8).

Territorial System

The relationship between central government and locale is reflected in two dimensions, national uniformity and provincial autonomy (Lieberthal & Oksenberg, 1990, p. 138).

There are four different levels of administration from top to bottom; these are parallel with the bureaucratic rank system within the vertical functional system. There are over 300 governments at the prefecture level, almost 3000 at the county level and over 40,000 at the township level (Martin, 2010, p. 15). Local government at and above county level has the authority of directing the work of governments at lower level (Chinanet, 2011). Due to its vertical management from top to down, information such as command and policy can be transmitted from central government to every unit of this country. Therefore, national uniformity is an important feature of the relationship between central leadership and local governments.

A local government has the authority to manage a wide range of physical, personnel and financial resources (Chinanet, 2011). As Martin (2010, p. 15) points out, a local government has, to some extent, autonomy to conduct authority. The continued reform, particularly the budgetary reform, provided the local government with an increased authority and more incentives to protect their own interests. The decrease in local dependence on central government leads to the rise of local autonomy.

However, as central government sought to maintain national uniformity, central policy might have been refused and delayed by local governments due to different preferences between central and local governments. In this sense, central decisions such as removing or appointing top leaders of local government played a significant role in controlling the expanding trend of provincial/local autonomy. Generally speaking, local officials had to implement the central policy. Meanwhile, they might damage interests of those officials in other functional bodies because of contradictory interests between the central policy and the local priority. Conversely, local officials did not implement the central policy in order to avoid conflicts with other functional departmental decisions.

Functional Systems

Functional systems refer to those ministries, commissions, committees, bureaus and state-controlled social organisations playing various roles in the national affairs (Saich, 2011). The highest administrative system is the State Council. In other words, the State Council is the top leader of functional systems.

Authority and communication in the functional system are delivered along two channels, vertical hierarchy inside the functional system and the chain of command inside the vertical territorial system. Therefore, this sort of governmental structure constrains efficient communication and cooperation between different functional systems (ministries).

A lack of collaboration between different ministries was a severe problem in the governing system (Christiansen & Rai, 1996). Due to their different preferences, several ministries had to confront frictions between them, and this could lead to the fragmentation of the governing system. As Lampton (1987) shows, in terms of the Three Gorges Project, the Ministry of Electric Power inclined to a small hydroelectric station which could be easily constructed, the Ministry of Finance was not favour of the expensive project which would require a huge financial budget, and the Ministry of Transportation warned that this project should ensure the safety of navigation.

Non-State Actors

The state transformation approach opens to the discussions on the role of non-state actors in the policy-making. Decentralisation and fragmentation not only exist in the territorial and functional administrative systems but also reflect the interaction between governmental bodies and non-state actors such as companies and non-governmental organisations.

For example, local environmental protection bureau has to face difficulties dealing with private entrepreneurs. These private entrepreneurs exert, to some degree, resistance and reluctance to the local bureau (Lo & Tang, 2006, pp. 205–206). The rise of public participation and civil society makes a great contribution to the transition in the environmental governance of China. Environmental non-governmental organisations play an important role in raising the public awareness of environmental issues and shaping the recognition of solutions and strategies (Xie, 2011, p. 207). Public participation in the Chinese political context can be achieved by framing and influencing the environmental debates and concerns such as the socially constructed green business (Martens, 2006, p. 211). The decentralisation and fragmentation of the governing system give a great space of environmental activities.

In addition to decentralisation and fragmentation, internationalisation has been identified in the transformation of the governing system in China. Foreign direct investments play an important role in the economic development of China. On the one hand, these foreign investments shift heavy-polluted industries from Western countries to China (Shi & Zhang, 2006, p. 275). This situation raises the environmental concern about local pollutions. On the other hand, these foreign investments have an incentive to improve local environmental regulations because the products exported to the Western markets must meet strict environmental standards (Jahiel, 2006, p. 325). In this sense, internationalisation plays an important role in improving the environmental governance of China (Xue et al., 2007). Also, China witnesses a wide participation of international non-governmental organisations, such as Greenpeace and World Wild Fund for Nature (WWF), in addressing environmental issues, and the international actors have raised the public awareness and concern about environmental qualities (Mol & Carter, 2006, pp. 160–161).

Therefore, the state transformation approach offers the concepts of decentralisation, fragmentation and internationalisation to understand the dynamic relations across functional, territorial and non-state actors. In the context of climate change issues, it is very important to interrogate to what extent various actors discursively emerge on the climate leadership in China.

Findings: A State Transformation Approach to Climate Leadership in China

Leadership on Climate Change at the Central Administrative Level

At the national level, the discourse on climate change is centralised around the Chinese government. The State Council of China is responsible for leading the climate change policy-making and identifying strategies and solutions. In terms of the climate change policy-making, the Chinese government identifies several important ministerial actors including the Ministry of Science and Technology, the National Development and Reform Commission, the Ministry of Environmental Protection, the

Ministry of Foreign Affairs, the Ministry of Finance and the State-owned Assets Supervision and Administration Commission (Hart et al., 2014, p. 8). Originally, the agency of representing China to attend the international climate institutions was the China Meteorological Administration (CMA). Due to the nature of climate-related science and research, the CMA participated in the works organised by the International Panel on Climate Change (IPCC) (Bjørkum, 2005). However, actions on addressing climate issues had been quickly redefined and reconstructed as economic development rather than pure scientific issues. In order to make a better coordination and cooperation between ministerial actors, the Chinese government set up the National Coordination Committee on Climate Change in 1998. This leading institution includes 15 ministerial bodies, and it is led by the National Development and Reform Commission which is the most important governmental actor in the process of the climate change policy-making (Bjørkum, 2005, p. 42). In 2007, this institution was replaced by the National Leading Group on Climate Change under the direct leadership of Premier Wen of the State Council (Tsang & Kolk, 2010, p. 192).

The institutional arrangements reflect the centralised discourse in the leadership on climate change policies. In June 2007, the National Development and Reform Commission released *the National Climate Change Programme* announcing the target of reducing energy intensity of China (NDRC, 2007). This programme states six principles of addressing climate change at the national level. They emphasise the importance of sustainable development, different historical responsibilities, mitigation and adaptation, the integration of policies, science and technology and international cooperation. On the one hand, these principles show positive Chinese attitudes towards addressing climate change. On the other hand, the focus on sustainable development and historical responsibilities reflects Chinese positions in international climate change negotiations. China positions itself as a developing country. Economic development remains a priority in political agenda although it should be achieved in a sustainable way. Also, being a developing country, China insists on the principle of common but differentiated responsibilities. This principle implies that while developed countries should have the historical responsibilities for addressing climate change and reducing emissions, developing countries should focus on economic development and poverty eradication. Developed countries have the responsibility for transferring technological and financial supports to developing countries. These principles form the fundamental positions in addressing climate change in China.

In 2014, the Chinese government issued *the China's National Plan on Climate Change for 2014–2020*. The discourse around principles changed from that in 2007 (NDRC, 2014, pp. 4–5). China's climate positions to some degree changed in 2014. The principle with an emphasis on domestic and international dimensions refers to a statement that China should be responsible for proactively participating in international climate change cooperation. The concepts of sustainable development and of common but differentiated responsibilities are not reiterated in the main principles of this important governmental document. These principles were not discarded but they were moved to the rest of the document.

More importantly, China has been rising substantially in the global climate leadership. The climate cooperation between China and the US makes a great contribution to the establishment of the 2015 Paris Climate Agreement (Schreurs, 2016). Also, China has been working closely with the EU reflected in the bilateral climate statements (Gippner & Torney, 2017).

It is important to note that the China's 2020 objectives of climate mitigation announced during the 2009 Copenhagen Summit had been achieved in 2018 (UNFCCC, 2018). In 2009, China promised to reduce carbon intensity by 40% to 45% compared with its level of 2005 in 2020 (NDRC, 2009). In fact, China's carbon intensity has been reduced 46% in 2017 (UNFCCC, 2018).

While the Chinese government controls its leadership at the national level in terms of climate change issues, various ideas and positions can be identified in the decentralised, fragmented and internationalised system across different central-local, vertical, horizontal and non-state actors. The leadership can be transformed and rearranged in the central-local, vertical and horizontal relations.

Vertical Rearrangement of Climate Leadership Across Central and Local Actors

The central-local relations are not like an absolute command system. A competition exists in the interaction between the central and provincial governments although it cannot be very transparent and fierce.

While the central government focuses on the national strategy of sustainable development and the historical responsibility of addressing climate change, local governments are concerned about their own issues such as energy saving. As Qi et al., (2008, pp. 393–395) explain, local governments prioritise economic growth as the top task in their political agenda. Energy supply and consumption provides a basis for securing economic growth. When energy is discursively linked with climate change, local governments are willing to make and improve the institutional arrangements. Clearly, the central governmental discourse about historical responsibility is not a primary concern at the local level.

The decentralisation between central and local governments raises a concern about the leadership on climate change. While the central leadership points to an ambitious goal of addressing climate change, the decentralised positions at the local level show various concerns and ideas. The sub-national actors have very different orientations towards the centralised discourse concerning climate change. This decentralisation raises these actors to share the leadership on climate and causes the institutional rearrangements at the local level. For example, the authority of setting provincial targets of reducing emissions had not been given to the provincial agencies until in 2006. This was mainly because provincial targets assigned by the national level were not achieved at the provincial level. In 2008, Shanxi Province, Inner Mongolia Autonomous Region and Jilin Province decreased the targets of emissions reductions originally set in the 11th Five-Year Plan while failing to achieve them (Qi, 2011b, pp. 29–33). This explicitly challenges the top-down approach and instead redistributes the leadership between central and provincial actors. However, this is not to undermine but enhance the leadership of China because local actors have incentive to raise their roles, interests, voices and positions in the climate governance.

Territorial and Functional Redistribution of Climate Leadership Between Local Governments and Environmental Agencies

The vertical system is located in the competitions between the territorial power and the functional body. On paper, the territorial power plays a leading role in its subordinary functional body. However, this relationship can be understood as bargaining through mobilising resources, motivating interests and controlling information and knowledge (Lieberthal &

Oksenberg, 1990, p. 159). The prominent example of this competition is the institutional competitions between local governments and environmental protection bureaus. Local governmental resistance to environmental regulations leads to a difficulty in the implementation of the climate change policy (Marks, 2010, p. 975). Environmental regulations implemented by the Environmental Protection Bureau conflict with economic growth prioritised by local government. Local government is concerned about its economic performance which might be influenced by punishing local heavy-polluted factories which make a great contribution to tax payments and a rapid GDP growth. On the one hand, the conflict lies on the relations between the environmental protection bureau and the local government. On the other hand, they can bridge the gap when the local government agrees with ideas from the environmental protection bureau encouraging business to invest in green and low carbon industry (Schröder, 2011, pp. 31–32). The green business can be understood as a common and shared idea between a local government and its environmental bureau in terms of the implementation of the climate policy.

This explanation with the competitions might be challenged by the argument that local governments have a higher administrative rank and have a greater power of mobilising materials and allocating financial resources than the environmental protection bureau at the local level (Xue et al., 2007). The administrative disparity between them is thus determined by the powers of material arrangements. However, this argument ignores a fact that economic development is discursively more popular than environmental protection. Obviously, economic priority is favoured not only by local governmental departments but also by companies. With the public awareness of environmental issues, the creation of the Ministry of Environmental Protection occurs as an important segment of the institutional reform in China (Qiu & Li, 2009). The rise of environmental discourse in the political agenda of China raises the capability of local environmental bureaus to compete with other governmental functional departments. Local government cannot ignore the pressures from the central governmental concern about climate change at the national level and from public anger over environmental pollution at the local level (Shi & Zhang, 2006, p. 288). If local governments want to address these concerns, they have to think about the choice of adjusting the institutional arrangements in a sustainable and environmental-friendly way. Also, with the emergence of the discourse of green economy, economic growth and environmental protection no longer oppose to each other. This shared green discourse leads to the improvement of institutional arrangements in terms of environmental governance. For example, the market-based mechanism became an important policy tool of climate change governance in China (Lo, 2015, p. 11).

Cross-Functional Redistribution of Climate Leadership: From NDRC to MEE

Various functional actors can reconstruct the climate leadership in China. Functional bodies at the ministerial level raise different concerns about addressing climate change. The centralised policy-making and various ideas among ministerial actors outline the prominent feature of the state transformation approach to China.

The National Development and Reform Commission is responsible for making national strategies of economic development, and it thus considers securing economic growth as the priority in the political agenda (Bjørkum, 2005, p. 43). An important solution to addressing climate change is to control the greenhouse gases emissions from human activities including industrial productions. Strict regulations over controlling emissions and pollutions can lead to negative effects on economic performance. Therefore, the primary concern of the National Development and Reform Commission is that the positive action on reducing emissions might pose a threat to the long-term economic development.

The Ministry of Foreign Affairs pays attention to sovereignty and China's cooperation with other developing countries in terms of addressing climate change (Bjørkum, 2005, pp. 43–44). It is one of the most important ministerial actors, and it is mainly responsible for coping with issues concerning international climate change negotiations (Chmutina, 2010b, p. 6). The concern about national sovereignty refers to a view that China is sensitive to limitations to its rights of utilising natural resources (Yu, 2008, p. 143; Harris, 2009, p. 61).

The Ministry of Science and Technology focuses on technological innovation and transfer and considers technology as an important solution to climate change (Bjørkum, 2005, p. 44). It plays a proactive role in encouraging low carbon development, technological innovation, economic transformation and industrial upgrade with a positive attitude towards mitigating climate change (Heggelund et al., 2010, p. 241; Stensdal, 2015, p. 58).

The Ministry of Environmental Protection plays a proactive role in addressing environmental issues and implementing climate change policies but it ranks as a marginal level in the policy-making process (Chmutina, 2010b, p. 6; Richerzhagen & Scholz, 2008). As Heggelund et al., (2010, p. 239) clarify, however, the rising environmental concern leads to the birth of the Ministry of Environmental Protection upgraded from the State Environmental Protection Administration.

Indeed, the environmental agency has witnessed a shift from a marginal role to a leader in the climate governance of China. China witnesses the bureaucratic upgrade from National Environmental Protection Agency (NEPA), the State Environmental Protection Administration (SEPA) to the Ministry of Environmental Protection (MEP) (Qiu & Li, 2009, p. 10,153). This bureaucratic upgrade shows that environmental protection has been rising in the political agenda of China. In March 2018, the Ministry of Ecology and Environment (MEE) was created particularly adopting the responsibility for addressing and mitigating climate change from the National Development and Reform Commission (NDRC). Also, it adopted the authorities of controlling agricultural pollution and managing underground water pollution from the Ministry of Agriculture and the Ministry of Land and Resources respectively (CCGov, 2018). Therefore, the MEE was emerging to be a leader in the administrative system of addressing climate issues in China.

However, the shift of the leadership from the NDRC to the MEE has potential institutional challenges. First, the competing priorities between economic development and environmental protection remain in the climate institutional framework (Qi & Wu, 2013). This is simply because economic development remains a primary task of China. In this sense, the NDRC continues to play an important role in the leadership on climate institutional framework. Second, as the national sovereignty and diplomacy is a focal point of the MoFA in the climate negotiations, to what extent the ambitious climate proposals of the MEE have been translated into practices requires a further observation.

In brief, these different functional bodies have different concerns and positions and thus reconstruct the climate leadership in a fragmented way. However, the institutional shifts in the leadership on climate change from the NDRC to the MEE demonstrate the China's proactive attitudes towards reducing carbon issues.

Internationalisation of Climate Leadership: Emerging Non-State Actors

The climate-related ideas and discourses, particularly like green growth and low carbon economy, have been internationalised and then adopted in the discursive context of climate governance of China. Having lacked of a structural power in the governing system of China, non-state actors cannot institutionally participate in the policy-making of China. However, they can support and discursively construct the notion of green development and enhance its important role in the climate leadership of China. Greenpeace (2016), being a famous international environmental NGO, identifies promoting green development as one of its primary tasks in the East Asia particularly including China. The main feature of these environmental NGOs is that they do not challenge the fundamental principles and polices of China. However, they have discursively been leading the social construction of environmental protection and emission protection as an important basis for a sustainable economic growth. Their discursive leadership has been demonstrated by ensuring that their discourse is acceptable and feasible to the Chinese government.

In addition to non-governmental organisations, academic actors play a leading role in socially constructing climate change through scientific publications and discussions emphasising the consequence of human activities and negative impacts over the ecological system (Heggelund et al., 2010, p. 240). The knowledge community attempts to challenge and even replace the principle of common but differentiated responsibilities with a support for a legally bind target of reducing emissions and appeals to the Chinese government to make a long-term policy of mitigating climate change (Wübbeke, 2011, p. 1015). Although this idea fails to change basic principles, it influences the official discourse of climate change. China promised that the carbon emissions would reach the peak by 2030 in 2014 (Stensdal, 2015, p. 52) and submitted its intended-national determined contribution proposal to the United Nation Framework Convention on Climate Change with the target of reducing carbon intensity in 2015 (Belis & Qi, 2015, pp. 199–200).

In 2002, the Stockholm Environment Institute (SEI) and the United Nations Development Programme (UNDP) released a report about the choice of green development for China. This report clarified environmental challenges and threats and raised the importance of industrial reforms and sustainable development in China (UNDP, 2002). Particularly, this report made a link between climate change and green development. In this sense, climate change could generate a serious effect on China's agricultural and ecological systems. Green development and reform was seen as an important means for mitigating climate issues (UNDP, 2002, p. 65). This demonstrates that international organisations and foreign academic institutions did contribute to the emergence of climate leadership by discursively raising environmental issues and economic development in China even in 2002.

The business actor plays an important role in linking economic opportunity with addressing climate change. On the one hand, the business actor considers mitigating climate change as an important feature of corporate social responsibility (CSR) (Jiang & Wong, 2016). The CSR had been emerging in the business communities in the West. On the other hand, low carbon economy has been slowly entering into the win–win discourse of addressing climate change (Qi & Wu, 2013, pp. 306–308). International financial institutions made a great contribution to the discursive adoption of green development in China. The World Bank considered green development an important opportunity. With the path of green development, China can secure a rapid economic growth without considerable carbon emissions and environmental damage, and it can achieve a sustainable development with investments in clean technologies, renewable energy and markets (WB, 2012, p. 233). Therefore, the ideas of CSR and green development have been adopted enhancing the climate governance of China, and thus the business actors have made a contribution to the share of the climate leadership in China.

In summary, while different actors have various interests and voices of climate issues, they have shared and redistributed the climate leadership by raising the ideas of green development and scientific knowledge. The internationalisation of climate leadership has been reflected in the process of adoption of the climate-related ideas in the context of climate governance of China.

A New Era of Climate Change Governance of China

In March 2018, China announced its ambitious reform on governmental structures. The institutional reform is designed to shift some governmental functions from the National Development and Reform

Commission (NDRC), which is seen as the most influential governmental body, to other ministerial government departments. The NDRC had played a key role in shaping climate change policies of China until March 2018. Its major role in making climate change policies was moved to the Ministry of Ecology and Environment which was created in March 2018 (CCGov, 2018). This means that an environmental government department started to play a leading role in making climate change policies of China. In this sense, China enters a new era of climate change governance raising environmental protection on its political agenda. The reform of environmental governance indicates a shift in climate policy towards raising environmental and ecological protection onto the agenda. This section also reviews the historical development of climate change governance and institutions of China and identifies possible weaknesses of the current institutional reform in China's climate change governance and politics.

Institutional Reform for Environmental Protection in China

In 1982, China created the Ministry of Urban and Rural Development and Environmental Protection which supervised the Environmental Protection Department (MEP, 2018). In 1988, the National Environmental Protection Agency (NEPA) was created and received a subministerial administrative rank. In 1998, it was renamed as the State Environmental Protection Administration (SEPA) and was upgraded to a ministerial rank (Carter & Mol, 2006, p. 152). A dramatic change in environmental governance can be demonstrated by the establishment of the Ministry of Environmental Protection (MEP) in 2008. However, as Qiu and Li (2009) stated, the MEP still confronted fundamental challenges from other ministries and encountered difficulties in collecting and sharing information with other governmental departments.

In order to improve the weak environmental governance, the Ministry of Ecology and Environment (MEE) gained environment-related administrative functions from other ministries in March 2018 (CCGov, 2018). Figure 2.1 shows that MEE adopted the functions from other ministerial bodies namely the National Development and Reform Commission (NDRC), the Ministry of Land and Resources (MLR), the Ministry of Water Resources (MWR), the Ministry of Agriculture (MOA), the State Oceanic Administration (SOA) and the South-to-North Water Diversion

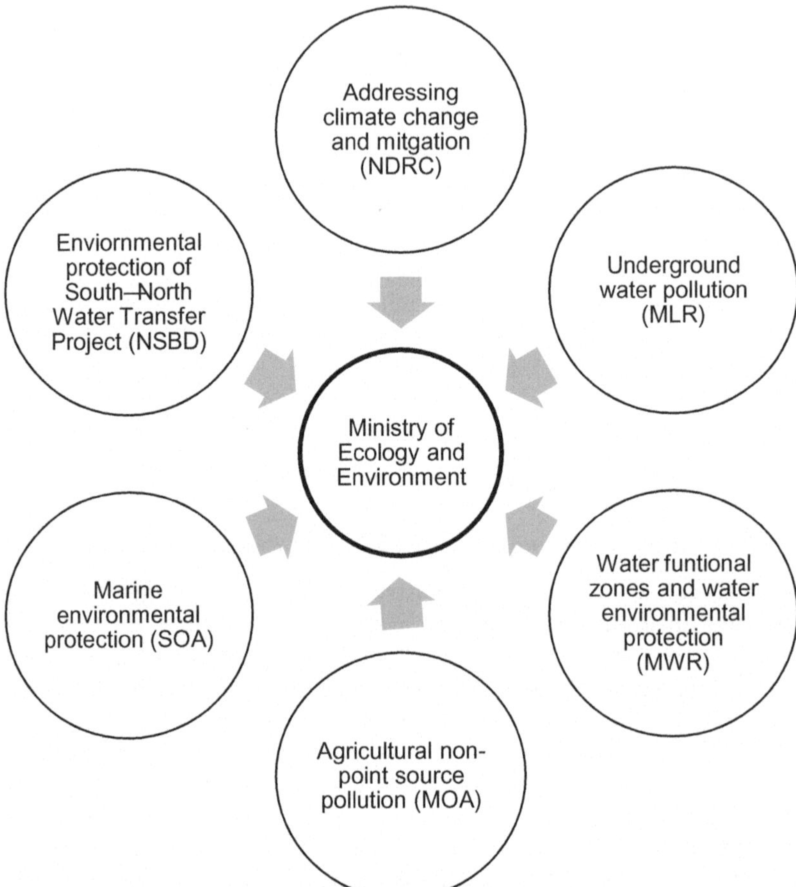

Fig. 2.1 Transferring environment-related functions from other ministries to the new Ministry of Ecology and Environment

Project Construction Committee (Nan Shui Bei Diao, NSBD). Therefore, water pollution policies may become more consistent because these functions will now be performed by a single entity.

The shifting of responsibility for addressing and mitigating climate change from the NDRC to the MEE is a substantial change in climate governance in China. Therefore, it is important to review the extent to

Table 2.1 Dynamic relationships between environmental agencies and climate change governance

Environmental agencies	*Addressing climate change*
NEPA (1988)	N/A
SEPA (1998)	A weak role
MEP (2008)	Controlling pollutants
MEE (2018)	Making climate policy

which environmental agencies address climate change in China (see Table 2.1).

Climate change issues had not entered the top Chinese political agenda until 1998 when China established the National Coordination Committee on Climate Change (Marks, 2010, p. 976). The SEPA played a weak role in climate change governance of China. This is mainly due to the fact that the SEPA had limited resources and institutional capacities compared to other ministerial bodies (Richerzhagen & Scholz, 2008). In 2008, the MEP was seen as an important governmental actor of climate change governance in China, with responsibility for controlling pollutants. However, because carbon dioxide was not identified as pollutants in China, the MEP was not mainly responsible for proposing measures against carbon dioxide emissions (Hart et al., 2014). In 2018, the MEE would be able to play a leading role in making climate change policies because it has shouldered the main responsibility for addressing climate issues from the NDRC.

A New Era of Climate Change Governance of China

Following the review on the development of environmental institutional designs of China, this section argues that Chinese climate change policy agenda enters a new era of environmental protection. The first phase of the climate governance of China focused on climate scientific research in the early 1990s. Due to its major role in conducting climate research, the China Meteorological Administration (CMA) was originally responsible for participating in the activities of the Intergovernmental Panel on Climate Change (IPCC) and leading the Chinese delegation to international climate negotiations (Bjørkum, 2005, p. 42). The second phase started in 1998 when the NDRC took the main responsibility for addressing climate change issues. It indicates that climate change policy agenda had been driven by economic factors during that phase.

The NDRC was seen as the most powerful governmental body and was responsible for making economic policies (Stensdal, 2014, p. 120). Also, with the rise of China's status in global governance, foreign policies had been increasingly linked to international climate negotiations. The Ministry of Foreign Affairs has been responsible for organising China's participation in international climate change conferences and its primary concerns are national sovereignty and diplomacy.

Notably, China enters the third phase of climate change governance raising environmental and ecological protection onto climate policy agenda. The MEE shouldering the main responsibility for addressing climate change indicates that Chinese climate policies will be built upon environmental criteria more than those for economic growth.

There are two main advantages of current institutional reform on climate change governance of China. First of all, the reform enhances consistency between climate change policy-making and implementation. The MEE can play a leading role in the climate change policy process. Secondly, the MEE adopting environmental institutional functions from other governmental bodies implies a dramatic rise in policy agenda for environmental protection. This can be beneficial to the efficiency and effectiveness of the shape and implementation of climate and environmental policies in China.

Dynamic Approach to the Climate Leadership at the Functional Level

The formation of the MEE can lead to coherence and consistency in climate change policy-making and implementation in China. However, it is very important to identify challenges to the institutional reform in environmental and climate change governance of China.

First of all, a competing relation between economic development and environmental protection remains in the new climate change governance. As a wide range of existing literature on climate governance of China demonstrates, economic development and priority plays a key role in climate change politics of China (Belis & Qi, 2015; Hatch, 2003; Lewis, 2007; Yu, 2008). Therefore, being responsible for making macroeconomic policy, the NDRC played a leading role in forming climate policy in China. However, the MEE shouldering the main responsibility for making climate policies does not mean that China prioritises climate issues on its political agenda. This is simply because economic development remains a core task of China. While the MEE can play a leading role in proposing measures against carbon emissions, it will have to consider the interests and policy ideas of other governmental bodies particularly like the NDRC. Even though the leading role of the NDRC in Chinese policy-making process declined due to the reform, it remains a key actor in the Chinese political system. Unless the top Chinese leaders fundamentally downgrade economic priority, the MEE will have to consider the importance of economic growth in its climate change policies.

Secondly, climate change policies have been closely linked to energy issues and considerations. Reducing emissions relies on controlling energy consumptions and improving energy mix (Green & Stern, 2016). China's energy mix heavily relies on coal consumption. Restricting coal consumption is seen as an effective solution to meeting emissions reduction. An obvious example is the introduction of harsh measures in November 2017 to ban coal consumption and encourage the use of natural gas (BBC, 2017). These measures can lead to an improvement of air quality and an achievement of emissions reduction. However, the MEE is not responsible for energy policies. It therefore has to address possible conflicts with other governmental bodies in the process of climate change policy-making.

Thirdly, local environmental agencies are constrained by local governments. For example, a municipal-level government is responsible for providing financial and human resources to its local environmental agency (Mol & Carter, 2006, p. 155; Schreurs, 2017, p. 169). If economic growth remains a primary task of the government, the environmental agency might confront a huge challenge over implementing an ambitious climate target at the local level.

Fourthly, the MEE is assumed to play a weak role in the climate change diplomacy of China. The Ministry of Foreign Affairs (MOFA) plays a key role in coordinating Chinese participation in international climate change negotiations. It has a strong stance on securing national interests in foreign affairs in terms of global climate politics (Chmutina, 2010a, p. 6). Even though the MEE institutionally receives more environmental governmental functions, it would not be as powerful as the NDRC and the MOFA.

It is not reasonable to expect China fundamentally to abandon economic growth. However, it is important to raise environmental protection and addressing climate change onto the Chinese political agenda. The establishment of the MEE reflects a step that raises environmental protection onto the political agenda. Therefore, the current reform on institutional arrangements indicates that Chinese climate change governance enters an era of ecological and environmental protection.

However, it is worth noting that the MEE has to confront the complicated challenges within current institutional arrangements. Therefore, this section offers three recommendations for strengthening the climate change governance of China. First of all, China should raise the status of the MEE in the National Leading Group on Climate Change. The NDRC had been mainly responsible for leading the group. This recommendation is designed to ease the tension between the MEE and other key governmental departments such as the NDRC. Secondly, enhancing the vertical management of the MEE can reduce the level of reliance of local environmental agencies on local governments. The vertical environmental governance can help achieve climate targets effectively, especially with a top-down approach to policy implementation. In March 2021, Huangrun Qiu, the Minister of Ecology and Environment, led a group to investigate the emissions of pollutants from the steel mills in Tanshan, Hebei Province. This shows the rising administrative power of the MEE in managing environment practices at local levels. Thirdly, the Minister of Ecology and Environment can be appointed as the top representative of the Chinese delegation to international climate change negotiations. Actually, Xie Zhenhua was appointed as the special climate envoy and was worked with the MEE in 2021. This appointment can be beneficial to China's desire to raise its leadership in global affairs, particularly international climate change governance.

Discussion and Conclusion

This research reveals the decentralised, fragmented and internationalised leadership on climate governance of China across a wide range of actors. The main argument is this transformed system does not weaken but enhances the climate leadership in China. It is very important to note that the decentralisation and fragmentation of climate discourse do not damage the climate governance but instead redistribute and share the climate leadership among the various actors.

First of all, the centralised leadership remains around the top level in China due to the administrative system. This is a fundamental structural power of ensuring a consistent policy-making and governance of addressing climate issues in China. The state transformation approach offers a dynamic process of understanding an evolution of climate governance of China. However, this does not mean a fundamental change in the system occurs. The sovereign states are the key units in the international climate negotiations. Therefore, the central leadership at the national level remains there and it is thus an important guarantee of the effective climate governance in China.

Secondly, the decentralisation of the climate leadership raises the local voices and concerns. While the climate policies can be made at the national level, their implementation and impacts take effect at the local level. Without the participation of local actors, climate policies could have not been effectively translated into practices. The decentralised system breaks the dichotomy of leaders at the national and followers at the local level. Local governmental agencies can participate in reconstructing the climate leadership. In this sense, on the one hand, they integrate their concern about economic impact into the considerations of improving climate governance. On the other hand, the impacts of climate change over local communities require attention from the provincial and national administrative levels. The multiple climate leadership implies the interactions between central and local levels.

Thirdly, the territorial and functional fragmentation of climate leadership has been identified and improved. A structural leadership matters in the context of governing system of China. Raising the administrative rank of the environmental agency at the national level helps improve the effectiveness of environmental governance. A higher administrative status particularly at the local level helps environmental officials overturn their marginal role in the leadership on addressing climate issues. With the institutional reform, the local environmental agency plays an important role in regulating industrial behaviour and fining local factors for their pollution and emissions. In addition to raising the rank, widening the functional tasks of the environmental agency explicitly raise its key role in the leadership on climate governance. While the NDRC remains the most powerful ministerial agency, the MEE plays a leading role in climate governance particularly after adopting the responsibility for forming climate policies. However, the structural leadership does not undermine the importance of other competing discourses. It is important to note that while the MEE moves to the centre of the stage, the fragmented system remains there. It is inevitable that other functional agencies potentially have different climate-relate ideas with the

MEE due to the various positions and interests in terms of climate and environmental governance.

Fourthly, the internationalisation determines the leadership on climate governance adopting various actors. While the various non-state actors do not play a leading role in climate governance, they have a potential to discursively raise the idea of green development in the climate leadership of China. The leadership on climate has not only been identified at the national level but it also has been shared by and redistributed among the academic, business, non-governmental actors at the international level. Particularly with the development of online information and social media, the relations among various leaders and followers have to be examined in a dynamic way (Gilani et al., 2020; O'Reilly et al., 2015).

Fifthly, this chapter employs the state transformation approach to mapping the leadership on climate governance in a decentralised, fragmented and internationalised system. A main contribution to the existing literature on climate leadership is to eliminate the boundary between the leaders and followers in the climate governance of China. The theoretical frameworks of leadership require a consideration of the context of China and thus have been examined with the state transformation approach. The redistribution of the climate leadership various actors has occurred within the governance of China. A discursive approach is employed to understand how climate-related discourses, particularly like economic development and environmental protection, have been interpreted across various actors. This can have a potential for improving the climate politics of China through diversifying, sharing and redistributing the climate leadership rather than simply raise a single actor to shoulder it.

Limitations and future research directions can be identified based on the above discussions. An empirical research is required to conduct in order to make a better understanding how specific actors play a leading/marginal role in the climate governance of China. Also, an analysis of a fit between the climate goals and their achievement requires a further study. More importantly, while the main findings of this chapter imply that the shared multiple leadership is beneficial to the effective climate governance of China, it is important to explain why other developing countries do not show this similar picture.

With the evolution of climate change politics of China, various actors will emerge and fade influencing policy options. It is very important to track how the decentralised, fragmented and internationalised process evolves and determines the leadership on climate change governance of

China. Also, it is very important to conduct further research on interactions between Chinese carbon governance and global climate politics. The future research agenda should observe how current environmental institutional reform can affect Chinese climate governance and diplomacy and global climate institutional arrangements.

References

Afionis, S., Fenton, A., & Paavola, J. (2012). EU climate leadership under test. *Nature Climate Change, 2*, 837–838.

Bach, M. (2019). The oil and gas sector: From climate laggard to climate leader? *Environmental Politics, 28*, 87–103.

BBC. (2017). China does U-turn on coal ban to avert heating crisis. *BBC*, 08/12/2017.

Belis, D., & Qi, Y. (2015). At the crossroads: China's domestic and international climate change policy. *Carbon & Climate Law Review, 199*.

Benson, D., Jordan, A., Cook, H., & Smith, L. (2013). Collaborative environmental governance: Are watershed partnerships swimming or are they sinking? *Land Use Policy, 30*, 748–757.

Biedenkopf, K., van Eynde, S., & Bachus, K. (2019). Environmental, climate and social leadership of small enterprises: Fairphone's step-by-step approach. *Environmental Politics, 28*, 43–63.

Bjørkum, I. (2005). *China in the international politics of climate change: A foreign policy analysis*. The Fridtjof Nansen Institute.

Carter, N. T., & Mol, A. P. J. (2006). China and the environment: Domestic and transnational dynamics of a future hegemon. *Environmental Politics, 15*, 330–344.

Case, P., Evans, L. S., Fabinyi, M., Cohen, P. J., Hicks, C. C., Prideaux, M., & Mills, D. J. (2015). Rethinking environmental leadership: The social construction of leaders and leadership in discourses of ecological crisis, development, and conservation. *Leadership, 11*, 396–423.

CCGov. (2018). *Project of institutional reform on state council (国务院机构改革方案)* [Online]. The State Council of the Chinese Central Government. http://www.gov.cn/guowuyuan/2018-03/17/content_5275116.htm [Accessed 01/04/2018].

Chan, N. (2018). "Large ocean states": Sovereignty, small islands, and marine protected areas in global oceans governance. *Global governance: A review of multilateralism and international organizations*, 24.

Chinanet. (2011). *China's political system* [Online]. http://www.china.org.cn/english/Political/25060.htm [Accessed 01/04/2018].

Chmutina, K. (2010a). China and climate change: The role of policy making in climate change mitigation. In *China Policy Institute*, T. U. O. N. (ed.). Nottingham.

Chmutina, K. (2010b). *China and climate change: The role of policy making in climate change mitigation.*

Christiansen, F., & Rai, S. (1996). *Chinese politics and society: An introduction.* Prentice Hall/Harvester Wheatsheaf.

Crossman, B., & Crossman, J. (2011). Conceptualising followership—A review of the literature. *Leadership, 7*, 481–497.

Crowley, K., & Nakamura, A. (2018). Defining regional climate leadership: Learning from comparative analysis in the Asia Pacific. *Journal of Comparative Policy Analysis: Research and Practice, 20*, 387–403.

de Águeda Corneloup, I., & Mol, A. P. J. (2014). Small island developing states and international climate change negotiations: The power of moral "leadership." *International Environmental Agreements: Politics, Law and Economics, 14*, 281–297.

Dupuis, J., & Schweizer, R. (2019). Climate pushers or symbolic leaders? The limits to corporate climate leadership by food retailers. *Environmental Politics, 28*, 64–86.

Eckersley, R. (2016). National identities, international roles, and the legitimation of climate leadership: Germany and Norway compared. *Environmental Politics, 25*, 180–201.

Eikeland, P. O., & Skjærseth, J. B. (2019). Oil and power industries' responses to EU emissions trading: Laggards or low-carbon leaders? *Environmental Politics, 28*, 104–124.

Ford, J., & Harding, N. (2015). Followers in leadership theory: Fiction, fantasy and illusion. *Leadership, 14*, 3–24.

Ford, J., & Harding, N. (2018). Followers in leadership theory: Fiction, fantasy and illusion. *Leadership, 14*, 3–24.

Gilani, P., Bolat, E., Nordberg, D., & Wilkin, C. (2020). Mirror, mirror on the wall: Shifting leader–follower power dynamics in a social media context. *Leadership, 16*, 343–363.

Gippner, O., & Torney, D. (2017). Shifting policy priorities in EU-China energy relations: Implications for Chinese energy investments in Europe. *Energy Policy, 101*, 649–658.

Green, F., & Stern, N. (2016). China's changing economy: Implications for its carbon dioxide emissions. *Climate Policy*, 1–15.

Greenpeace. (2016). *Greenpeace exists because this fragile earth deserves a voice.* [Online]. http://www.greenpeace.org/eastasia/about/ [Accessed 03/03/2019].

Gupta, J., & Grubb, M. (2000). Climate change and European leadership. In J. Gupta & M. Grubb (Eds.), *Climate change and European leadership*. Kluwer.

Hajer, M. A. (1995). *The politics of environmental discourse: Ecological modernization and the policy process*. Oxford University Press.

Hameiri, S., & Jones, L. (2015). Rising powers and state transformation: The case of China. *European Journal of International Relations, 22*, 72–98.

Hameiri, S., Jones, L., & Heathershaw, J. (2019). Reframing the rising powers debate: State transformation and foreign policy. *Third World Quarterly, 40*, 1397–1414.

Hameiri, S., & Zeng, J. (2019). State transformation and China's engagement in global governance: The case of nuclear technologies. *The Pacific Review*, 1–31.

Harris, P. G. (2009). *Climate change and foreign policy: Case studies from east to west*. Taylor & Francis.

Harris, P. G., Chow, A. S., & Karlsson, R. (2013). China and climate justice: Moving beyond statism. *International Environmental Agreements: Politics, Law and Economics, 13*, 291–305.

Hart, C., Zhu, J., Ying, J., Cassisa, C., & Kater, H. (2014). *Mapping China's climate policy formation process*. China Carbon Forum.

Hatch, M. T. (2003). Chinese politics, energy policy, and the international climate change negotiations. In P. G. Harris (Ed.) *Global warming and East Asia: The domestic and international politics of climate change*. Routledge.

Heggelund, G. (2004). *Environment and resettlement politics in China: The three gorges project*. Ashgate.

Heggelund, G., Andresen, S., & Buan, I. F. (2010). Chinese climate policy: Domestic priorities, foreign policy and emerging implementation. *Global commons, domestic decisions: The comparative politics of climate change*, 229–259.

Held, D., Nag, E. M., & Roger, C. (2011). The governance of climate change in China. *Preliminary Report, LSE-AFD Climate Governance Programme*.

Hensengerth, O. (2014). *Between local and global norms: Hydropower policy reform in china*. Springer.

Hensengerth, O. (2015). Multi-level governance of hydropower in China? The problem of transplanting a western concept into the Chinese governance context. *Multi-Level Governance: The Missing Linkages*.

Hjerpe, M., Linn¨¦R, B. O., Parker, C., & Karlsson, C. (2011). Looking for leaders: Perceptions of climate change leadership among climate change negotiation participants. *Global Environmental Politics, 11*, 89–107.

Jahiel, A. R. (2006). China, the WTO, and implications for the environment. *Environmental Politics, 15*, 310–329.

Jänicke, M., & Wurzel, R. K. W. (2019). Leadership and lesson-drawing in the European Union's multilevel climate governance system. *Environmental Politics, 28*, 22–42.

Jiang, W., & Wong, J. K. (2016). Key activity areas of corporate social responsibility (CSR) in the construction industry: A study of China. *Journal of Cleaner Production, 113*, 850–860.

Kanie, N. (2003). *Assessing leadership potential for beyond 2012 climate change negotiation: Elaborating a framework of analysis.*

Karlsson, C., Hjerpe, M., Parker, C., & Linnér, B.-O. (2012). The legitimacy of leadership in international climate change negotiations. *Ambio, 41*, 46–55.

Karlsson, C., Parker, C., Hjerpe, M., & Linnér, B. R.-O. (2011). Looking for leaders: Perceptions of climate change leadership among climate change negotiation Participants. *Global Environmental Politics, 11*, 89–107.

Lampton, D. M. (1987). Chinese politics: The bargaining treadmill. *Issues and Studies, 23*, 11–41.

Learmonth, M., & Morrell, K. (2016). Is critical leadership studies 'critical'? *Leadership, 13*, 257–271.

Learmonth, M., & Morrell, K. (2017). Is critical leadership studies 'critical'? *Leadership, 13*, 257–271.

Lewis, J. I. (2007). China's strategic priorities in international climate change negotiations. *The Washington Quarterly, 31*, 155–174.

Lieberthal, K., & Oksenberg, M. (1990). *Policy making in China: Leaders, structures, and processes*. Princeton University Press.

Lieberthal, K. G. (1992). Introduction: The 'fragmented authoritarianism'model and its limitations. In K. Lieberthal, & D. M. Lampton (Eds.), *Bureaucracy, politics, and decision-making in Post-Mao China, studies on China*. University of California Press.

Lindenthal, A. (2014). Aviation and climate protection: EU leadership within the International Civil Aviation Organization. *Environmental Politics, 23*, 1064–1081.

Lo, A. (2015). *Carbon trading in China: Environmental discourse and politics.* Palgrave Macmillan.

Lo, C. W.-H., & Tang, S.-Y. (2006). Institutional reform, economic changes, and local environmental management in China: The case of Guangdong province. *Environmental Politics, 15*, 190–210.

Mai, Q., & Francesch-Huidobro, M. (2014). *Climate change governance in Chinese cities*. Taylor & Francis.

Malnes, R. (1995). 'Leader' and 'entrepreneur' in international negotiations: A conceptual analysis. *European Journal of International Relations, 1*, 87–112.

Marks, D. (2010). China's climate change policy process: Improved but still weak and fragmented. *Journal of Contemporary China, 19*, 971–986.

Martens, S. (2006). Public participation with Chinese characteristics: Citizen consumers in China's environmental management. *Environmental Politics, 15*, 211–230.

Martin, M. F. (2010). *Understanding China's political system*. DTIC Document.

Mazmanian, D. A., Jurewitz, J. L., & Nelson, H. T. (2020). State leadership in U.S. climate change and energy policy: The California experience. *The Journal of Environment & Development, 29*, 51–74.

Meijerink, S., & Stiller, S. (2013). What kind of leadership do we need for climate adaptation? A framework for analyzing leadership objectives, functions, and tasks in climate change adaptation. *Environment and Planning C*, 31.

MEP. (2018). *The history of sturcture* [Online]. Ministry of Environmental Protection of China. http://english.sepa.gov.cn/About_MEE/History/ [Accessed 04/01/2018].

Mol, A. P. J., & Carter, N. T. (2006). China's environmental governance in transition. *Environmental Politics, 15*, 149–170.

NDRC. (2007). *China's national climate change programme*. National Development and Reform Commissions.

NDRC. (2009). *China announces targets on carbon dioxide emission cuts* [Online]. Department of Climate Change, National Developmentand Reform Commission. http://en.ccchina.gov.cn/Detail.aspx?newsId=38858&TId=123 [Accessed 23/01/2015].

NDRC. (2014). *The notification of National Plan on climate change for 2014–2020*. National Development and Reform Commission.

Newig, J., & Fritsch, O. (2009). Environmental governance: Participatory, multi-level–and effective? *Environmental Policy and Governance, 19*, 197–214.

O'reilly, D., Leitch, C. M., Harrison, R. T., & Lamprou, E. (2015). Leadership, authority and crisis: Reflections and future directions. *Leadership, 11*, 489–499.

Ongaro, E. (2015). *Multi-Level governance: The missing linkages*. Emerald Group Publishing Limited.

Ostrom, E. (2012). Nested externalities and polycentric institutions: Must we wait for global solutions to climate change before taking actions at other scales? *Economic Theory, 49*, 353–369.

Palacková, E. (2017). The race for climate leadership in the ERA of trump and multilevel governance. *European View, 16*, 251–260.

Parker, C. F., & Karlsson, C. (2010). Climate change and the european union's leadership moment: An inconvenient truth? *Journal of Common Market Studies, 48*, 923–943.

Parker, C. F., & Karlsson, C. (2018). The UN climate change negotiations and the role of the United States: Assessing American leadership from Copenhagen to Paris. *Environmental Politics, 27*, 519–540.

Piattoni, S. (2010). *The theory of multi-level governance: Conceptual, empirical, and normative challenges*. Oxford University Press.

Qi, X. (2011). The rise of basic in UN climate change negotiations. *South African Journal of International Affairs, 18*, 295–318.

Qi, Y. (2011). *Annual review of low-carbon development in China 2010*. Science Press.

Qi, Y., Li, M. A., Huanbo, Z., & Huimin, L. (2008). Translating a global issue into local priority. *The Journal of Environment & Development, 17*, 379–400.

Qi, Y., & Wu, T. (2013). The politics of climate change in China. *Wiley Interdisciplinary Reviews: Climate Change, 4*, 301–313.

Qiu, X., & Li, H. (2009). China's environmental super ministry reform: Background, challenges, and the future. *Environmental Law Reporter*.

Richerzhagen, C., & Scholz, I. (2008). China's capacities for mitigating climate change. *World Development, 36*, 308–324.

Sabatier, P. A., Focht, W., Lubell, M., Trachtenberg, Z., Vedlitz, A., & Matlock, M. (2005). Collaborative approaches to watershed management. In F. Sabatier, T. Lubell, & M. Vedlitz (Eds.), *Swimming upstream: Collaborative approaches to watershed management* . Massachusetts Institute of Technology.

Saich, T. (2011). *Governance and politics of China*. Palgrave Macmillan.

Sartori, G. (1970). Concept misformation in comparative politics. *American Political Science Review, 64*, 1033–1053.

Schreurs, M. (2017). Multi-level climate governance in China. *Environmental Policy and Governance, 27*, 163–174.

Schreurs, M. A. (2010). Multi-level governance and global Climate change in East Asia. *Asian Economic Policy Review, 5*, 88–105.

Schreurs, M. A. (2016). The Paris climate agreement and the three largest emitters: China, the United States, and the European Union. *Politics and Governance, 4*, 219–223.

Schröder, M. (2011). *Local climate governance in China: Hybrid actors and market mechanisms*. Palgrave Macmillan.

Shi, H., & Zhang, L. (2006). China's environmental governance of rapid industrialisation. *Environmental Politics, 15*, 271–292.

Stensdal, I. (2014). Chinese climate-change policy, 1988–2013: Moving on up. *Asian Perspective, 38*, 111–135.

Stensdal, I. (2015). China: Every day is a winding road. In G. Bang, G., A. Underdal, & S. Andresen (Eds.), *The domestic Politics of global climate change: Key actors in international climate cooperation*. Edward Elgar.

Tobin, P. (2017). Leaders and laggards: Climate policy ambition in developed states. *Global Environmental Politics, 17*, 28–47.

Torney, D. (2015). *European climate leadership in question*. The MIT Press.

Torney, D. (2019). Follow the leader? Conceptualising the relationship between leaders and followers in polycentric climate governance. *Environmental Politics, 28*, 167–186.

Tsang, S., & Kolk, A. (2010). The evolution of Chinese policies and governance structures on environment, energy and climate. *Environmental Policy and Governance, 20*, 180–196.

UNDP. (2002). Making green development a choice. *China Human Development Report 2002*. UNDP.

UNFCCC. (2018). *China meets 2020 carbon target three years ahead of schedule* [Online]. https://unfccc.int/news/china-meets-2020-carbon-target-three-years-ahead-of-schedule [Accessed 01/10/2021].

Van Vugt, M., & Von Rueden, C. R. (2020). From genes to minds to cultures: Evolutionary approaches to leadership. *The Leadership Quarterly,* 31, 101404.

WB. (2012). Seizing the opportunity of green development in China. *China 2030*. World Bank.

Wübbeke, J. (2011). The power of advice: Experts in Chinese climate change politics. *Fridtjof Nansen Institute (FNI) Report,* 15, 60.

Wurzel, R. K. W., Liefferink, D., & Torney, D. (2019). Pioneers, leaders and followers in multilevel and polycentric climate governance. *Environmental Politics, 28*, 1–21.

Wurzel, R. K. W., Moulton, J. F. G., Osthorst, W., Mederake, L., Deutz, P., & Jonas, A. E. G. (2019). Climate pioneership and leadership in structurally disadvantaged maritime port cities. *Environmental Politics, 28*, 146–166.

Xie, L. (2011). China's environmental activism in the age of globalization. *Asian Politics & Policy, 3*, 207–224.

Xue, L., Simonis, U. E., & Dudek, D. J. (2007). Environmental governance for China: Major recommendations of a task force. *Environmental Politics, 16*, 669–676.

Yu, H. (2008). *Global warming and China's environmental diplomacy*. Nova Science Pub Inc.

Zhang, B. (2019). State transformation goes nuclear: Chinese national nuclear companies' expansion into Europe. *Third World Quarterly, 40*, 1459–1478.

Zhang, Y., & Zheng, Y. (2008). New development in China's climate change policy. *China House Discussion Paper,* 30.

Zürn, M., Wälti, S., & Enderlein, H. (2010). *Introduction*. Edward Elgar.

CHAPTER 3

The Rise of China in the Global Governance of Climate Change: From 2009 Copenhagen Summit to 2015 Paris Negotiations

Abstract Climate change is seen as one of significant global crises. China's contribution and participation is a key to global success in addressing climate change. International pressures and rapid growth of economy and emissions could be understood as important drivers of a change in China's climate change policies and positions. However, they may exert influences over climate policy through discourse. For this reason, this work raises the central research question: How do domestic political discourses shape China's resolution on the leadership of global governance of climate change? This chapter argues that the change in domestic political discourse makes a great contribution to China's positions on climate change. Research methods involve the discourse analysis of policy documents in 2009 and 2015.

Keywords Climate change · China · Copenhagen · Paris Agreement · Governance · Discourse

Climate change is seen as one of significant global crises. China's contribution and participation is a key to global success in addressing climate change. International pressures and rapid growth of economy and emissions could be understood as important drivers of a change in China's climate change policies and positions. However, they may exert

S. Wang, *Climate Change Discourse in China*,
https://doi.org/10.1007/978-981-16-6754-1_3

influences over climate policy through discourse. For this reason, this work raises the central research question: How do domestic political discourses shape China's resolution on the leadership of global governance of climate change? This chapter argues that the change in domestic political discourse makes a great contribution to China's positions on climate change. Research methods involve the discourse analysis of policy documents in 2009 and 2015.

The International Panel on Climate Change (IPCC) had released five scientific reports from 1990 to 2014. The scientific community has confirmed the causal link between human activities and global warming. The greenhouse gases emissions since the industrial revolution make a great contribution to the cause of climate change (IPCC, 2014). Therefore, a key solution to addressing climate change is mitigating the emissions. International society has made efforts on constructing international climate change institutions. The milestones of global climate action had been demonstrated by the creation of the United Nations Framework Convention on Climate Change (UNFCCC) in 1992 and the Kyoto Protocol in 1997 (Bolin, 2007).

The existing international climate change institutions place an emphasis on the principle of common but differentiated responsibilities. According to the principle, while developed countries should undertake the historical responsibility for addressing climate change and reducing emissions, developing countries have the rights of economic development and poverty eradication and can acquire the financial and technological transfer from developing countries (Cullet, 2010; Rajamani, 2000; Yu & Zhu, 2015). China is categorised as the group of developing countries who do not have the obligations of accepting the legally binding compulsory targets of emissions reduction in international agreements (Harris et al., 2013).

Since the UNFCCC and the Kyoto Protocol came into effect, the world has witnessed the rapid economic growth of major developing countries such as China, Brazil and India over 20 years. More importantly, these countries are recognised as emerging economies and have produced a big amount of greenhouse gases emissions (Hochstetler & Viola, 2012; Hurrell & Sengupta, 2012; Stensdal, 2015). China even surpassed the US to become the largest global emitter since 2005 (WB, 2016b). Therefore, it is very important to discuss the role of China in international climate change governance and cooperation.

This chapter focuses on the relationship between domestic political discourse on the rise of China and its participation in international climate change negotiations. The main argument of this research is the discourse on the rise of China affects its positions on global climate change governance. Therefore, it identifies policy discourses on the rise of China and international climate change cooperation in 2009 and 2015. The Copenhagen climate change summit and the Paris conference were held in 2009 and 2015, respectively. The two events reflect the milestones of global climate change governance (Dimitrov, 2016).

Firstly, this chapter briefly introduces China's environmental governance and global climate change politics. Secondly, it discusses the importance of understanding climate change discourses. Thirdly, it demonstrates the research methods employed in this research. *The Annual Report on the Work of the Government* is used to understand the policy discourse on China's role in international affairs and governance. The annual report '*China's Policies and Actions for Addressing Climate Change*' is selected to analyse shifts in China's positions on international climate change negotiations. Fourthly, it looks at the domestic political discourse on the rise of China in international governance. Lastly, this chapter discusses how dynamic China's positions on global climate change governance reflect the evolution of its attitudes towards international cooperation.

China's Environmental Governance and Global Climate Change Politics

China witnesses the rise of environmental issues on its political agenda. While economic development remains a significant consideration in national affairs, environmental protection and ecological conservation have been adopted in national economic and social policies particularly like Five-Year Plan (NDRC, 2007). Mitigating carbon emissions is considered a core solution to addressing climate change. It is in this sense that mitigation can be beneficial to environmental protection and addressing climate change. Therefore, this section discusses how China's environmental governance has been adopted into global climate change politics.

The Rise of China's Environmental and Climate Governance

China's environmental governance was initiated by China's participation in the United Nations Conference on the Human Environment held in Stockholm in 1972 (Economy, 2010). At the conference, China had chances to communicate with other countries about environmental problems and information concerning China and the world. However, China believed that it should not be responsible for environmental degradation caused by industrialised Western countries (Economy, 2010). It is reasonable to assume that China did not adopt environmental protection into its core political claims and positions particularly in the 1970s. China's economic level was far behind that in developed and industrialised countries. Environmental degradation caused by economic development was thus not seen as a primary concern.

However, following the international conference, the first National Conference on Environmental Protection was organised by the State Council of China in 1973. Environmental problems started to enter political discussions at the central level of China. Institutionally, in 1984, China created the State Environmental Protection Commission which is responsible for managing and regulating environmental issues (Qiu & Li, 2009), and it set environmental protection as a national basic policy and was engaged in preventing pollution (Mol & Carter, 2006). In 1988, in order to raise institutional capability, China launched the National Environmental Protection Agency (NEPA). While China attempted to improve its environmental governance, the NEPA remained a weak role in raising environmental issues on China's political agenda which is dominated by economic priority (Hatch, 2003). In 1998, it was upgraded to become the State Environmental Protection Administration. However, this institutional arrangement failed to make it to have greater voice in China's policy-making system, and it received its ministerial status with the creation of the Ministry of Environmental Protection in 2008 (Qiu & Li, 2009).

While environmental protection had a weaker status than economic development on national agenda, the Chinese government had not marginalised environmental and climate issues. In 1990, China organised the National Climate Change Coordination Leading Small Group preparing for attending the 1992 Rio Earth Summit. Being Chinese representative to the IPCC, the China Meteorological Administration

(CMA) was mainly responsible for coordinating climate change governance in China (Zhang & Zheng, 2008). In 1998, the Chinese government realised that climate change issues had involved a wide range of economic and social problems beyond scientific research and it thus created the National Coordination Committee on Climate Change linking different ministries. The leading role in coordinating climate change governance shifted from the CMA to the NDRC which plays an important role in making national economic policies (Marks, 2010). This cross-ministerial political arrangement can be beneficial to policy coordination among different governmental actors in terms of China's positions in international climate change negotiations (Bjørkum, 2005). In 2007, the National Leading Group on Climate Change was created under the direct leadership of Premier Wen of the State Council. It was responsible for making domestic policies of emissions reduction and coordinating positions on international climate change conferences (Tsang & Kolk, 2010). Also, China released for the first time the objectives of addressing climate change in *the National Climate Change Programme* and indicated that China would reduce 20% energy intensity by 2010. This reflects one milestone of China's climate change governance as addressing the issues entered China's political agenda (Stensdal, 2014).

China's Participation in Global Climate Change Politics

In addition to domestic climate change governance, participation in international negotiations is another important dimension of China's response to climate change issues. According to the principle of common but differentiated responsibilities under the United Nations Framework Convention on Climate Change and the Kyoto Protocol, China was categorised into the group of developing countries. While developing countries have the rights of economic development and poverty eradication, developed countries should be historically responsibility for mitigating emissions and providing financial and technological support for developing countries (Marks, 2010). Therefore, economic development and common but differentiated responsibilities are two fundamental principles of addressing climate change of China (Zhang, 2003; Held et al., 2011).

However, it is reasonable to believe that China plays a key role in global climate change politics. Firstly, China is the largest greenhouse gases emitter and a leading economy (Harris, 2010). Secondly, China's per-capita emissions reached the average level of the world (WRI, 2014). In this sense, China's responsibility for addressing climate change has been emerging and rising. Without China's substantial participation on international climate change governance, global collective action on mitigating emissions cannot be successful (Harris et al., 2013). China has to bear greater international pressure and expectation. In order to show a positive attitude towards attending the 2009 Copenhagen climate change summit, China announced the target of reducing 40 to 45% carbon intensity by 2020. Particularly, Premier Wen Jiabao attended the summit on behalf of the Chinese government (Stensdal, 2014). The dramatic shift in China's role in international climate change negotiations has been witnessed due to the leadership of President Xi especially since 2013. During the Beijing Asia-Pacific Economic Cooperation (APEC) summit, China and the US made a joint statement on addressing climate change issues in November 2014 (Schreurs, 2016). China unprecedentedly made a promise to achieve a carbon peak around 2030. More importantly, China submitted its Intended Nationally Determined Contributions to the UNFCCC in June 2015, signed the Paris Climate Change Agreement in December 2015 and ratified it in September 2016 (WRI, 2016).

Therefore, this chapter focuses on exploring how the rise of China in global governance has discursively influenced international climate change negotiations. It looks at the discursive links between the rise of China in global affairs and its positions on climate change issues.

Climate Change Discourse

Discourse can be seen as an important perspective of understanding climate change politics of China. There are two key features of discourse deployed in this chapter. Firstly, discourse can affect political arrangements and provide policy options (Hajer & Versteeg, 2005; Hajer & Wytske, 2013). In this sense, it is very important to understand how climate change discourses influence the evolution of China's attitudes towards and positions on global climate change governance. While China has discursively linked economic development to addressing climate change issues, it emphasises economic priority (Held et al., 2011). Also, China discursively constructs it as a developing country which does not have the historical responsibility for substantial emissions action (Heggelund et al., 2010). Therefore, economic development and historical responsibility had been two fundamental principles of the Chinese government in climate change discourses. China's position on the historical responsibility is in accordance with the principle of common but differentiated responsibilities under the existing international climate change institutions. According to the UNFCCC and the Kyoto Protocol, China is categorised as the party who does not have the obligation of taking compulsory action on mitigating emissions (Christoff, 2010; UN, 2014). Rather, China has the rights of acquiring financial and technological support from developed countries. Therefore, the principle of common but differentiated responsibilities is a fundamental position of China in international climate change negotiations.

Secondly, discourse can be dynamic influencing the development of climate change policy options and arrangements. For example, in the early stage of international climate change negotiations, China was reluctant to agree with the Clean Development Mechanism (CDM) because it was concerned that industrialised countries would utilise this policy arrangement to shirk their obligations of reducing emissions and shift their responsibilities for addressing climate change to developing countries (Bjørkum, 2005). The discursive link between CDM and concern about climate justice demonstrates why China showed its disagreement with this arrangement. However, China changed its attitude towards and adopted the CDM when it was discursively linked with positive economic, financial and technological benefits. Since 2000, China had started to raise the level of approving the CDM projects. China realised that the adoption of the CDM could help strengthen its capability to develop clean and renewable energy and technology (Harris & Yu, 2005).

Therefore, it is very important to study how the dynamic discourses have influenced China's role in global climate change governance. Not surprisingly, one might argue that China's role in climate change negations is determined by external factor, particularly like international pressures. Indeed, it is reasonable to assume that with a sharp increase in the levels of economic growth and carbon emissions of China, the international society necessarily exerts pressures on China. However, the international pressures should not be taken for granted. It is very important to understand how the international society has socially and discursively constructed China's role in global climate change governance and its obligations of and responsibilities for emissions reduction. This work does not focus on what pressures and expectations many countries have placed on China. Rather, the key to understanding China's role in global climate politics is analysing how the pressures and expectations have discursively affected China's climate change policy options and ideas. They might have a contribution to determine China's position on international climate change negotiations. Therefore, this chapter looks at how China understands its role in addressing climate change issues over time.

Also, the core contribution of this chapter is analysing and understanding how domestic political discourse concerning China's role in global governance and international cooperation makes a great contribution to China's participation in international climate change negotiations particularly the 2009 Copenhagen summit and the 2015 Paris conference. The Chinese government raised the importance of the two conferences on its political agenda. It can be demonstrated that Premier Wen attended the Copenhagen summit and President Xi delivered a strong and positive message at the Paris conference (Xinhuanet, 2016). However, the fundamental difference between these two conferences is that China showed a very positive position on achieving an international agreement and making a promise to reach a carbon peak at the Paris conference (Green & Stern, 2016). Therefore, it is very useful to compare the China's discourses on global governance and cooperation and climate change positions between 2009 and 2015. This chapter argues that the discourse on China's role in international governance makes a contribution to its adjustments to positions on global collective action on mitigation.

Method

As the discourse on China's role in global climate governance is a core contribution, this chapter employs discourse analysis to understand how China discursively adjusts its position on international climate change negotiations. It discusses the texts around China's resolution over raising its voice and status in global governance and existing international institutions. Also, it analyses the discursive link between the rise of China and its proactive participation in global climate governance. In addition to texts, social contexts are discussed in terms of climate change issues. This chapter takes various climate-related concerns such as economic development into consideration. In addition to climate change discourses, China's attitudes towards global governance and cooperation are seen as a key to understanding the climate change politics of China.

Firstly, this research identifies policy documents as one data source using the Annual Reports on the Work of the Government in 2009 and in 2015. This identification is due to the 2009 Copenhagen climate summit and the 2015 Paris conference. The Annual Reports are delivered by the State Council to the National People's Congress. They reflect the annual review on governmental work and clarify annual plans and principles for next year (Saich, 2016). This type of policy documents is identified and selected mainly because it can be used to explore China's attitudes towards proactive participation in international regimes.

Secondly, in order to understand China's governmental position on international climate change governance and cooperation, this research identifies *China's Policies and Actions for Addressing Climate Change—The Progress Report* in 2009 and in 2015, respectively, as key climate policy documents. This report shows China's attitude towards international climate cooperation with other countries (NDRC, 2016).

The Rise of China in Global Governance and International Cooperation

An important perspective of understanding the rise of China in global climate change politics is observing the discursive shifts in its participation in existing international system and order. It is very interesting to know how the domestic political discourse on international governance affects China's positions on climate change negotiations. This research identifies the Annual Reports on the Work of Government delivered by the State Council in 2009 and 2015, respectively. It is used to find how the Chinese government defines China's diplomatic relations and international cooperation.

As the statement below illuminates, the Chinese government expressed its positive attitudes towards international cooperation especially after the 2008 global financial crisis. The main point of the statement is China's concern about international financial arrangements. It is reasonable to assume that China saw addressing financial crisis as an important opportunity for raising its international status and reputation. This is mainly because China plays a key role in driving global economic growth. In 2009, while the global market was shrinking, China still witnessed a rapid economic growth rate reaching 9.4% (WB, 2016a). The economic performance showed a leading role of China in global economic recovery. Therefore, by utilising the opportunity of an unprecedented financial crisis, China was engaged in changing the role of developing countries in international financial systems and raising the voice of emerging economies such as China and India in existing institutions such as the World Bank and International Monetary Fund (IMF) (Ferdinand & Wang, 2013).

> We will increase friendly contacts with other countries in the world to create a favourable external environment for the steady and rapid development of China's economy. We will continue to deepen pragmatic cooperation with other parties, work with them to curb the spread of the global financial crisis, promote reform of the international financial system, and oppose trade and investment protectionism, in order to help bring about an early recovery of the world economy. (StateCouncil, 2009)

There are three features of the statement clarified in the 2009 Annual Report. Firstly, it simply identified other parties as the subject of international cooperation. The statement did not specify the exact type of partier such as developing countries for international cooperation. This demonstrates that China had not developed a very specific target for participating in global governance. Secondly, China would continue the pragmatic cooperation with other countries. This statement did not show a very ambitious attitude towards China's role in global governance. Thirdly, the statement demonstrated that China held a very positive position on the reform of the existing international financial system.

On the one hand, China was engaged in playing an important role in improving global financial governance. This demonstrates China's focus on international economic and financial affairs. Also, it means that it has been interested in participating in the reform and improvement of global governance since the 2008 financial crisis. On the other hand, it did not show a very obvious and positive position in involving other international systems particularly like global climate change institutions.

However, as the statement below demonstrates, the 2015 Annual Report reflects a shift in China's policy discourse on international governance and cooperation. In November 2014, under the leadership of President Xi, the Asia-Pacific Economic Cooperation (APEC) summit was held in Beijing showing China's ambitious desire to be a leading role in global affairs (Dimitrov, 2016).

> We will help develop a new type of international relations based on mutual benefit and cooperation, work to deepen strategic dialogue and practical cooperation with other major countries, and work to build a sound and

> stable framework for major-country relations. We will work comprehensively to make progress in neighbourhood diplomacy and work to create a community with a common future with our neighbours. We will work to strengthen unity and cooperation with other developing countries and protect our common interests. We will actively participate in international and multilateral affairs and work to make both the international system and the international order more just and equitable. (StateCouncil, 2015)'

Firstly, the Chinese government clarified three types of subjects as the target of developing and constructing international cooperation. These subjects are other major countries, neighbours and other developing countries. Based on the context of the statement, the major countries mainly refer to major economies particularly the members of Group of Twenty (G20). The neighbours refer to those countries geographically near China. This demonstrates the rising China's concern about its relations with neighbours. Other developing countries refer to a wide range of developing countries across Africa, Asia and Latin America.

Secondly, based on the three types of countries, China developed its definitions of various diplomatic relations. In 2015, China identified the major countries in international order and was engaged in constructing a stable relation with them. This shows China's positive position on involving international affairs through working with other key stakeholders of international society. Also, China developed 'a community with a shared future' as a discursive framework of constructing diplomatic relations with its neighbours. This demonstrates that China's neighbouring diplomacy became very important. While China held a very positive attitude towards other major countries and raised the importance of its neighbours, it vowed to secure common interests among developing countries. This demonstrates that China still categorised itself in the group of developing countries and played a leading role among them.

Thirdly, the 2015 policy discourse showed an obvious shift in China's position on participating in international governance. China expressed a very clear attitude towards actively participating in international affairs. Also, China made a resolution to participate in adjusting and improving international system and order.

As Table 3.1 shows, there are obvious adjustments to China's positions on international participation, cooperation and governance. This research finds two fundamental features of the adjustments. One is the categorisation of three types of diplomatic relations based on the different nature

Table 3.1 China's positions on international participation and cooperation in 2009 and 2015

Annual Report elements	*2009*	*2015*
Subject	Other parties	Other major countries; neighbours; other developing countries
Diplomatic relations	Pragmatic cooperation	Strategic dialogue and practical cooperation; a sound and stable framework for major-country relations; a community with a common future; our common interests
International governance	Promote reform of the international financial system	Actively participate in international and multilateral affairs; make both the international system and the international order more just and equitable

of countries. China developed specific corresponding measures of dealing with different types of countries. It is reasonable to assume that China could adjust its positions on international climate change negotiations and governance based on the categorisation. Another one is a very clear attitude towards participating in international affairs and improving international systems. Based on this position, it is very important to observe and understand how China would actively participate in global collective action on mitigating emissions and be engaged in improving existing international climate change governance in a just and equitable way. Therefore, the next section discusses how China's role in global climate change affairs has been discursively constructed.

China's Positions on International Climate Change Cooperation

There are two perspectives of understanding China's role in international climate change negotiations. Firstly, this research looks at China's official discourse on its role in climate change issues in 2009 and 2015. Secondly,

Table 3.2 China's position on international climate change governance and cooperation in 2009 and 2015

Year *Subject*	*2009*	*2015*
The subjects of international climate change cooperation	Related countries and regions	Developed countries
	The United Nations, international organisations, and foreign research institutions	International organisations
	Other developing countries	Developing countries South-South cooperation

it is very important to explore how other actors had socially constructed the role of China in the international negotiations.

By analysing the statements derived from China's Policies and Actions for Addressing Climate Change in 2009 and in 2015 respective, this research identifies the evolution of China's position on international climate change governance and cooperation. As Table 3.2 shows, China had adjusted the subjects of international climate change cooperation between 2009 and 2015.

Generally speaking, the similarity of the statements between 2009 and 2015 reflects China's focus on the cooperation with international organisations and developing countries. However, this research identifies three main adjustments to China's positions on international climate change cooperation.

Firstly, China recognised developed countries as one subject of international climate change cooperation in 2015. The 2015 statement showed a very positive attitude towards cooperation with other major parties including the US and the EU. While China did not abandon its position on the historical responsibilities of industrialised nations for reducing emissions, it expressed a strong desire to strengthen cooperation with developed countries in 2015.

Secondly, while China narrowly focused on international cooperation on climate-related research projects in 2009, it emphasised the importance of cooperation with a broad range of international organisation with various projects including green and low carbon economy in 2015. The subject of international climate cooperation was extended beyond the

United Nations and research institutions to a wider range of organisations such as the World Bank.

Thirdly, on the one hand, China still emphasised its cooperation with other developing countries. On the other hand, there were changes on China's attitude towards developing countries in 2015. China mentioned the cooperation among emerging economies particularly like BASIC countries (Brazil, South Africa, India and China). While these countries are labelled as developing countries under the UNFCCC and the Kyoto Protocol, they were major economies with rapid economic growth and major emitters with sharp increase in carbon emissions. They were therefore called for taking greater responsibilities for reducing emissions and substantially participating in international climate change governance. Also, China discursively raised the importance of South-South Cooperation in the policy document and vowed to provide support for other developing countries to address climate change issues. In this sense, China specified its different types of international cooperation between major emerging economies and developing countries vulnerable to climate change in 2015.

Therefore, compared to the statements in 2009, the China's policy discourse in 2015 showed a very positive attitude towards global climate change cooperation specifying the corresponding measures of dealing with developed countries, various international organisations, emerging economies and vulnerable nations.

Discursive Interaction Between the Rise of China and Global Climate Change Governance

Discourse can affect political arrangements, provide policy options and interact with other discourses (Hajer & Wytske, 2013; Hajer & Versteeg, 2005). It is in this sense that the domestic political discourse on the rise of China has affected the emergence of China's proactive participation in global climate change governance. China's shift in policy discourse on diplomatic relations reflects its discursive adjustments in international climate change cooperation. Conversely, international pressures and expectations make a significant contribution to the reproduction of the discourse on China's role in global governance and international cooperation. Especially since the 2008 global financial crisis, China has witnessed its rising status in international society. The rise of China can be demonstrated by its leadership in BRICS (Brazil, Russia, India, China and South Africa) and BASIC and its key role in APEC and G20.

First of all, the domestic political discourse on the rise of China affects China's role in global climate change governance. As Table 3.1 demonstrates, it is very obvious that the 2015 policy discourse on China's diplomacy showed more specific subjects of international cooperation and more proactive attitudes towards participating in and improving international system and order. On the one hand, China started to emphasise the importance of diplomatic relations with other major countries. This reflects China's proactive positions on cooperation with other major economies in international climate change negotiations in 2015. On the other hand, China still recognised itself as a developing country and focused on securing the common interests among developing countries. This can be demonstrated by China's fundamental principle of common but differentiated responsibilities in terms of climate change issues. While developing countries particularly those major emitters should take action on controlling emissions, developed countries have the historical responsibilities for addressing climate change issues and providing technological and financial support for developing countries. Under the discursive context of proactive participation in international system, China shows a positive shift in improving international climate change governance.

Secondly, climate change issues have been a key topic around global governance. Based on China's desire to play a more proactive role in international affairs, global climate change politics and governance can be understood as an important opportunity for China to improve its national image and reputation. China has made a great resolution on conserving energy and reducing emissions and pursues the achievement of sustainable development (Stensdal, 2015). The domestic demand of economic transformation and industrial upgrade makes a great contribution to China's economic, energy and climate change policies. Controlling energy consumption and carbon emissions had discursively been emerging and particularly adopted in the 11th Five-Year Plan of China (Green & Stern, 2016). China has taken substantial action on making domestic policies on reducing emissions particularly since the release of *the 2007 National Climate Change Programme* (Marks, 2010). In this sense, with the rise of public awareness of climate change issues, China has improved its national reputation and image particularly around international climate change governance. The prominent example is a common effort on addressing climate change between the US and China. In 2014, President Obama and President Xi announced a joint statement on climate change issues vowing to reduce greenhouse gases emissions. Particularly, they expressed a clear sign of achieving the 2015 Paris Climate Change Agreement (Dimitrov, 2016). The international expectations provide an opportunity for China to show its positive image and transform its domestic discourse on energy conservation and emissions reduction into international commitments.

Thirdly, international pressures influence domestic political discourse on the rise of China. There are four main aspects of international pressures on China's climate diplomacy. Firstly, China has become the largest greenhouse gases emitter. Without China's substantial action on emissions reduction, global collective action on addressing climate change cannot be successful (Harris, 2010). International society exerts a huge pressure on China to adjust its positions on international climate change negotiations. Secondly, the per-capita emissions of China reached the average level of the world (WRI, 2014). China could not utilise low emissions to defend its climate positions. China was therefore called for undertaking greater responsibility for reducing emissions. Thirdly, China has become a major economy and has more or less financial capability to mitigate and adapt to climate change. It thus was required to undertake international obligations and responsibilities providing support to other developing countries vulnerable to climate change. Fourthly, China has played an important role in international affairs. International society raises a pressure on China to play a leading role in global climate change governance. Therefore, international pressures can influence the domestic political discourse on the rise of China in international system and order. In 2015, China showed a very positive position on participating in and improving international system and order. It is in this sense that China has been shifting its role in global climate change governance from a learner to a proactive actor.

Is China a Leader of Global Climate Change Governance?

China can be seen as a key role in leading the existing climate change governance. China has signed and ratified the Paris Climate Change Agreement promising to reach a carbon peak by 2030 and controlling the consumption of coal. China's desire to become a leader in global climate change politics can be affected by its domestic discourses on environmental protection and international status. With the rise of public awareness and concern about environmental issues particularly like air pollution, China is engaged in taking substantial action on controlling emissions (McMullen-Laird et al., 2015). The domestic environmental discourse affects a positive China's attitude towards playing a leading role in climate change negotiations. Also, China makes efforts on raising its status in international system. Because China is a major economy and the largest greenhouse gases emitter, international climate change governance is a great platform to show its positive image.

However, China could not discursively abandon social development and historical responsibility as the fundamental principle of participating in international climate change politics. Firstly, economic development remains the important task on the political agenda of China. Even though economic growth has been slower, economic development will carry on discursively dominating the Chinese politics. While international society exerts a huge pressure on China to reduce emissions, it remains highly expected to seeing China as a leading role in global economic recovery. Secondly, it is reasonable to assume that China cannot abandon the principle of common but differentiated responsibilities in international climate change negotiations (Wübbeke, 2013). Based on the principle, China can call developed countries to transfer financial and technological support to developing countries including China. More importantly, this principle makes a fundamental difference in responsibilities for addressing climate change between developed and developing countries. If China still recognises itself as a developing country, it will not abandon the two fundamental principles.

It is very interesting if China will undertake the global leadership of existing international climate change governance after the Brexit and the US Trump administration. Undoubtedly, the EU plays a leading role in global climate politics ranging from the Rio Earth Summit, the Kyoto Protocol, and the Copenhagen summit to the Paris Climate Change Agreement (Schreurs, 2016). However, with the Brexit and other future uncertainties, the influence of the EU will be shrinking. The US had witnessed its withdrawal from the Paris Agreement during the Trump Administration. China will carry on playing a key role in international climate change governance based on its domestic discourse on environmental concerns and international cooperation. It is reasonable to assume that China will integrate its domestic demands of economic development, energy strategy and environmental protection into the international climate change negotiations.

Conclusion

This chapter argues that the domestic political discourse on the rise of China in international cooperation makes a significant contribution to Chinese attitudes towards global climate change governance. It shows an obvious adjustment to the role of China in international governance and cooperation from 2009 to 2015. Also, it analyses how the evolution of China's positions in international climate change negotiations reflects the discourse on the role of China in international cooperation.

China has shown more specific diplomatic measures of making cooperation with major countries and developing countries in 2015 compared to those in 2009. Based on the framework of the policy discourse, China developed different corresponding diplomatic relations with developed countries, emerging economies and developing countries vulnerable to climate change in the 2015 climate change positions. Also, China showed a more positive attitude towards contributing to global climate change governance with other major emitters and providing financial support to other developing countries. Therefore, this chapter concludes that the domestic political discourse can affect China's political arrangements of international climate change cooperation.

However, there are two important recommendations for future research. Firstly, it is very important to observe how China has discursively constructed its climate diplomatic relations with developing countries and how it differentiates emerging economies and vulnerable nations. Secondly, it is very interesting to expect how China will see the improvements of existing climate change institutions. This research establishes an important basis for exploring the future global climate politics.

References

Bjørkum, I. (2005). *China in the international politics of climate change: A foreign policy analysis*. The Fridtjof Nansen Institute.

Bolin, B. (2007). *A history of the science and politics of climate change: The role of the intergovernmental panel on climate change*.

Christoff, P. (2010). Cold climate in Copenhagen: China and the United States at COP15. *Environmental Politics, 19*, 637–656.

Cullet, P. (2010). Common but differentiated responsibilities. *Research Handbook on International Environmental Law*, 161.

Dimitrov, R. S. (2016). The Paris agreement on climate change: Behind closed doors. *Global Environmental Politics*.

Economy, E. C. (2010). *The river runs black: the environmental challenge to China's future*. Cornell University Press.

Ferdinand, P., & Wang, J. (2013). China and the IMF: From mimicry towards pragmatic international institutional pluralism. *International Affairs, 89*, 895–910.

Foot, R., & Walter, A. (2010). *China, the United States, and global order*. Cambridge University Press.

Green, F., & Stern, N. (2016). China's changing economy: implications for its carbon dioxide emissions. *Climate Policy*, 1-15.

Hajer, M., & Versteeg, W. (2005). A decade of discourse analysis of environmental politics: Achievements, challenges, perspectives. *Journal of Environmental Policy & Planning, 7*, 175–184.

Hajer, M. & Wytske, V. (2013). Voices of vulnerability: The reconfiguration of policy discourses. In J. S. Dryzek, B. N. Richard, & S. David (Eds.), *The Oxford handbook of climate change and society*. Oxford University Press.

Harris, P. G. (2010). *China and climate change: From Copenhagen to Cancun*.

Harris, P. G., Chow, A. S., & Karlsson, R. (2013). China and climate justice: Moving beyond statism. *International Environmental Agreements: Politics, Law and Economics, 13*, 291–305.

Harris, P. G., & Yu, H. (2005). Environmental change and the Asia Pacific: China responds to global warming. *Global Change, Peace & Security, 17*, 45–58.

Hatch, M. T. (2003). Chinese politics, energy policy, and the international climate change negotiations. In *Global warming and East Asia: The domestic and international politics of climate change*, 43¨C65.

Heggelund, G., Andresen, S., & Buan, I. F. (2010). Chinese climate policy: Domestic priorities, foreign policy and emerging implementation. In *Global commons, domestic decisions: The comparative politics of climate change* (pp. 229-259).

Held, D., Nag, E. M. & Roger, C. (2011, January). *The governance of climate change in China preliminary Report*, LSE-AFD Climate Governance Programme.

Hochstetler, K., & Viola, E. (2012). Brazil and the politics of climate change: beyond the global commons. *Environmental Politics, 21*, 753–771.

Hurrell, A., & Sengupta, S. (2012). Emerging powers, North-South relations and global climate politics. *International Affairs, 88*, 463–484.

IPCC. (2014). Climate change 2014 synthesis report summary for policymakers. *The Fifth Assessment Report of the Intergovernmental Panel on Climate Change*. Intergovernmental Panel on Climate Change.

Marks, D. (2010). China's climate change policy process: Improved but still weak and fragmented. *Journal of Contemporary China, 19*, 971–986.

Martin, M. F. (2010). *Understanding China's political system*. DTIC Document.

Mcmullen-Laird, L., Zhao, X., Gong, M., & Mcmullen, S. J. (2015). Air pollution governance as a driver of recent climate policies in China. *CCLR*, 243.

Mol, A. P. J., & Carter, N. T. (2006). China's environmental governance in transition. *Environmental Politics, 15*, 149–170.

NDRC. (2007). *China's national climate change programme*. National Development and Reform Commissions.

NDRC. (2016). *China's policies and actions on climate change*. The National Development and Reform Commission.

Qiu, X., & Li, H. (2009). China's environmental super ministry reform: Background, challenges, and the future. *Environmental Law Reporter*.

Rajamani, L. (2000). The principle of common but differentiated responsibility and the balance of commitments under the climate regime. *Review of European Community & International Environmental Law, 9*, 120–131.

Saich, T. (2016). State-society relations in the people's Republic of China post-1949. *Brill Research Perspectives in Governance and Public Policy in China, 1*, 1–57.

Schreurs, M. A. (2016). The Paris climate agreement and the three largest emitters: China, the United States, and the European Union. *Politics and Governance, 4*, 219–223.

StateCouncil. (2009). *Report on the work of the government*. The State Council of China.

StateCouncil. (2015). *Report on the work of the government*. The State Council of China.

Stensdal, I. (2014). Chinese climate-change policy, 1988–2013: Moving on up. *Asian Perspective, 38*, 111–135.

Stensdal, I. (2015). China: Every day is a winding road. In *The domestic politics of global climate change: Key actors in international climate cooperation*, 49.

Tsang, S., & Kolk, A. (2010). The evolution of Chinese policies and governance structures on environment, energy and climate. *Environmental Policy and Governance, 20*, 180–196.

UN. (2014). *First steps to a safer future: Introducing the United Nations framework convention on climate change* [Online]. http://unfccc.int/essential_background/convention/items/6036.php [Accessed 20/11/2014].

WB. (2016a). *Data* [Online]. The World Bank. http://data.worldbank.org/indicator/NY.GDP.MKTP.KD.ZG?end=2015&locations=CN&name_desc=false&start=2000&view=chart [Accessed 13/01/2017].

WB. (2016b). *Total greenhouse gas emissions* [Online]. The World Bank. [Accessed 05/01/2019].

WRI. (2014). *6 graphs explain the world's top 10 emitters* [Online]. World Resources Institute. http://www.wri.org/blog/2014/11/6-graphs-explain-world%E2%80%99s-top-10-emitters [Accessed 07/10/2016].

WRI. (2016). *Statement: India joins Paris agreement on climate change* [Online]. World Resources Institute. http://www.wri.org/news/2016/10/statement-india-joins-paris-agreement-climate-change [Accessed 16/10/2016].

Wübbeke, J. (2013). China's climate change expert community—Principles, mechanisms and influence. *Journal of Contemporary China, 22*, 712–731.

Xinhuanet. (2016, January 10). Xi Jinping's three ideas of green development: Improving people's lives, promoting economic development and delivering a promise (习近平绿色发展三大思路: 绿色惠民、绿色富国、绿色承诺). *Xinhuanet*.

Yu, H.-Y., & Zhu, S.-L. (2015). Toward Paris: China and climate change negotiations. *Advances in Climate Change Research, 6*, 56–66.

Zhang, Y. Z., & Yongnian (2008). *New development in China's climate change policy*. China House University of Nottingham.

Zhang, Z. (2003). The forces behind China's climate change policy. In *Global warming and East Asia: The domestic and international politics of climate change* (pp. 66-85). Routledge.

CHAPTER 4

Discursive Evolution Around the Principle of Common but Differentiated Responsibilities in China

Abstract Since the 1992 United Nations Framework Convention on Climate Change (UNFCCC) and the 1997 Kyoto Protocol, international society has witnessed the establishment of global climate change regimes and governance. The principle of common but differentiated responsibilities is seen as an important basis for international cooperation and governance. However, there are various interpretations of the principle. China puts an emphasis on the principle as a basic position on international climate change negotiations. This chapter discusses how the discourse on climate change responsibilities has evolved from 2007 to 2015 in China. A central argument is that the discursive change affects China's policy rhetoric and positions on international climate change governance. It can be used to explain China's proactive action on climate issues such as its ratification of the Paris Climate Change Agreement.

Keyword Climate change · Common but differentiated responsibilities · Discourse coalitions · Discourse network analysis · China

Since the 1992 UNFCCC and the 1997 Kyoto Protocol, international society has witnessed the establishment of global climate change regimes and governance. The principle of common but differentiated responsibilities is seen as an important basis for international cooperation and

S. Wang, *Climate Change Discourse in China*,
https://doi.org/10.1007/978-981-16-6754-1_4

governance. However, there are various interpretations of the principle. China puts an emphasis on the principle as a basic position on international climate change negotiations. This chapter discusses how the discourse on climate change responsibilities has evolved from 2007 to 2015 in China. A central argument is that the discursive change affects China's policy rhetoric and positions on international climate change governance. It can be used to explain China's proactive action on climate issues such as its ratification of the Paris Climate Change Agreement.

Climate change is mainly caused by human activities particularly greenhouse gases emissions from industrial production and consumption (IPCC, 2014). This determines international society to take global collective action on reducing emissions. Also, climate change can generate a wide range of negative effects on people's wellbeing threating survival and security. This means that the international society has a common responsibility for addressing climate change issues across national borders. On the other hand, national responsibilities for and capabilities to addressing climate change vary among different countries (Harris, 2009, p. 80). Human activities and emissions since the industrial revolution have been a main cause of the rising global average temperatures recently observed (IPCC, 2014). In this sense, industrialised countries should have the historical responsibilities for addressing climate change. They also have advanced technologies and financial capabilities in terms of climate change mitigation and adaptation. Compared to the industrialised nations, developing countries should not have the responsibilities for and obligations of addressing climate change with the low levels of national historical and per-capita emissions. While the developing countries have the weak levels of economic development and financial capabilities, they have to concern about their vulnerability to climate change. Therefore, the international society should have the different responsibilities for addressing the issues with the consideration of climate change justice.

The principle of common but differentiated responsibilities has been adopted in international climate change governance and institutions particularly the UNFCCC and the Kyoto Protocol. However, while the parties to the UNFCCC have a common agreement with climate change justice, they have different interpretations of the principle (Winkler & Rajamani, 2014, p. 107).

China sees the principle as a fundamental position on participating in international climate change governance and cooperation. According to the principle, China was categorised as a developing country and it thus did not have the obligations of taking the compulsory legally binding target of emissions reduction (Bjørkum, 2005, p. 28). However, other major stakeholders in international climate change negotiations challenge the China's fundamental position. For example, the Bush administration of the US quitted the Kyoto Protocol and called the major developing countries such as China and India to undertake the responsibilities for addressing climate change (Bang, 2015, p. 160).

This chapter sets a key research question: how the discourse on the principle of common but differentiated responsibilities evolves in China. In order to address the question, it employs a social constructionist approach to understand how climate discourse evolves and influences China's adjustments to positions in climate change negotiations.

Firstly, this chapter reviews dynamic and various interpretations of the principle of common but differentiated responsibilities in the context of global climate change politics. Secondly, it discusses the role of discourse in analysing China's positions on different responsibilities for addressing climate change. Thirdly, it identifies the main discourses on the responsibilities in China. The research uncovers how the discourse has been dominant and evolved. It discusses the discursive adoption in climate change policy rhetoric of China. Finally, the discussion section focuses on how the discursive change affects China's participation in international climate change negotiations.

The Principle of Common But Differentiated Responsibilities

The first IPCC report was issued in 1990 and it provided scientific evidence for the 1992 Rio Earth Summit in which the UNFCCC was created (Giddens, 2009, p. 185). The UNFCCC shows the principle of common but differentiated responsibilities particularly concerning the historical responsibility of addressing climate change. However, different countries have various understandings and interpretations of the responsibilities (Friman, 2013, p. 223).

Harris (2009, p. 92) demonstrates that sovereign states are the subject of undertaking the responsibilities for addressing issues in terms of international climate change justice. In this sense, countries are categorised into different groups for mitigating climate change. There are two statements below cited from the UNFCCC document. They show different responsibilities among developed and developing countries

> The Parties should protect the climate system for the benefit of present and future generations of humankind, on the basis of equity and in accordance with their common but differentiated responsibilities and respective capabilities. Accordingly, the developed country Parties should take the lead in combating climate change and the adverse effects thereof. (UN, 1992, p. 4)

As the statement above shows, according to the principle of common but differentiated responsibilities and respective capabilities, developed countries should play a leading role in addressing climate change. The categorisation of the difference rests on responsibilities and capabilities (Winkler & Rajamani, 2014, p. 105). Developed countries should be responsible for their historical emissions and total national emissions while developing countries favour this principle as they do not have compulsory obligations of substantial emissions reduction (Harris & Symons, 2013, p. 19). Also, the levels of per-capita in developing countries are much lower than those in industrialised nations. In terms of mitigation and adaption, developed countries have financial and technological capabilities to address climate change.

> The specific needs and special circumstances of developing country Parties, especially those that are particularly vulnerable to the adverse effects of climate change, and of those Parties, especially developing country Parties, that would have to bear a disproportionate or abnormal burden under the Convention, should be given full consideration. (UN, 1992, p. 4)

In addition to responsibilities and capabilities of developed countries, the statement above illuminates that developing countries have their national circumstances and some of them are vulnerable to the negative effects of climate change. Firstly, in terms of international climate change justice, developing countries are seen as the victims of climate change issues mainly caused by developed countries. In this sense, developed countries should have the historical responsibilities for helping developing countries mitigate the adverse effects of climate change. Secondly, due to the different levels of capabilities, developed countries should be responsible for providing financial and technological support to developing countries in terms of climate change mitigation and adaptation. Thirdly, developing countries require survival emissions and have their priority as economic development and poverty eradication. In other words, according to the principle of common but differentiated responsibilities, developing countries have the rights of achieving social and economic development.

While developed countries are categorised into the group of stakeholders who should have the responsibilities for addressing climate change, they have different positions in international negotiations. The prominent feature is the gap between the EU and the US (Blaxekjær & Nielsen, 2015, p. 752).

The EU plays a proactive and leading role in global climate change negotiations and governance. It has established the EU Emissions Trading Scheme (ETS) in order to utilise economic tools for controlling carbon emissions. Also, it announced ambitious targets of addressing climate change. For example, the EU issued the 2030 Climate and Energy Framework in October 2014 cutting greenhouse gases emissions at least 40%, increasing the share of renewable energy to at least 27% of total consumption and improving energy efficiency to 27%.

While the Clinton administration signed the Kyoto Protocol, President Bush did not support the ratification of it in the US in 2001. The main reason for the US withdrawal from the Kyoto Protocol was the lack of compulsory mitigation targets of major developing countries (Bang, 2015, p. 160). The US was reluctant to accept an international agreement without substantial participation of emerging economies particularly like China (Foot & Walter, 2010, p. 200). However, the US has shifted its positions on international climate negotiations during the Obama administration. The prominent feature of the US climate diplomacy is its cooperation with China in a wide range of fields including renewable energy, technology and green industry (Giddens, 2009, pp. 221–222).

In order to respond to the US withdrawal from the Kyoto Protocol, developing countries particularly like China could not accept the institutional arrangements of international climate change governance without the US commitments to emissions reduction (Heggelund, 2007, p. 177). In the early stage of climate change negotiations, developing countries created a coalition namely G-77 securing their basic interests and emphasising different historical responsibilities and national capabilities. Being fundamentally different to developed countries, these developing countries focused on their own low levels of per-capita emissions and the large historical and current emissions of developed countries (Brunnée & Streck, 2013, p. 592). Even in 2009, the G-77 and China emphasised the importance of historical responsibility in international climate change governance and agreements (Friman, 2013, p. 226). Also, the levels of per-capita emissions are very low among the developing countries. Therefore, these countries can prioritise economic and social development and poverty eradication.

However, the social, economic and political contexts of the world have changed, and some developing countries such as China, India and Brazil are seen as emerging economies with the sharp growth of economic development and carbon emissions (Winkler & Rajamani, 2014, pp. 599–600). China surpassed the US to become the largest greenhouse gases emitter after 2005 (WB, 2016) and its per-capita emissions were nearly that of the EU and were much higher than that of other emerging economies such as India even in 2011 (WRI, 2014). Particularly since the 2008 global financial crisis, the world has witnessed that emerging economies such as China and India play an important role in international economic affairs and global financial governance (Ferdinand & Wang, 2013, p. 895). Therefore, the major developing countries are seen as the stakeholders who should be responsible for and have capabilities of addressing climate change. As Harris et al., (2013, p. 293) state, without the substantial participation of China, the global collective action on stabilising climate system cannot be achieved successfully.

China has adjusted its positions on international climate change negotiations undertaking more responsibilities (Belis et al., 2015, p. 211; Schreurs, 2016, p. 222). On the one hand, China remains a strong supporter of existing international climate change institutions under the leadership of the UN and emphasises the fundamental principle of common but differentiated responsibilities (Harris et al., 2013, pp. 299–300; Wübbeke, 2013, p. 712). On the other hand, China's discourses on different responsibilities between developed and developing countries have evolved. Firstly, China has shifted its role in global climate issues from blaming the US for inaction to proactively making cooperation with the US. Secondly, China recognises the responsibilities of emerging economies and major emitters for addressing climate change while it has defended the importance of economic development according to the principle of common but differentiated responsibilities. Thirdly, it appears that other developing countries vulnerable to climate change, such as the Alliance of Small Island States (AOSIS), are highly expected to seeing the greater responsibilities of emerging economies such as China for reducing emissions (Stensdal, 2015, p. 53). China has to review and adjust its role in climate concerns raised by other developing countries (Yu & Zhu,

2015). Therefore, it is very important to understand how the different responsibilities have been socially constructed in China.

China's Discourse on the Responsibilities for Addressing Climate Change

Generally speaking, like the divided interpretations of the principle of common but differentiated responsibilities, China's discourse on the responsibilities for addressing climate change reflects two different dimensions. A primary discourse refers to an emphasis on the fundamental principle of common but differentiated responsibilities and the existing international climate change institutions particularly under the framework of the UN. It includes criticisms over the US weak action on reducing emissions. The US has been socially constructed as a key emitter who shirks its responsibility for addressing climate change issues. This discourse indicates that it is unacceptable to force developing countries to take substantial action without the leading role of the US in reducing emissions. Also, according to the fundamental principle, the developed countries should be responsible for reducing emissions and providing technological and financial support to developing countries.

On the other hand, there is an emerging 'common responsibility' discourse indicating positive attitudes towards taking substantial action on climate change mitigation and adaptation. Firstly, major economies and greenhouse gases emitters such as the US and China have been discursively constructed to be a leading role in commitments to emissions reduction. Secondly, developing countries particularly like emerging economies such as India, Brazil and China have been discursively constructed to undertake the responsibilities for addressing climate change. Thirdly, China is seen as a key economy, a major emitter and an important actor in international affairs and global governance. It thus had been understood to be responsible for holding a proactive position on international climate change negotiations.

There are two different discourses on the climate change justice and responsibilities. The *different responsibilities* discourse refers to the responsibility of developed countries, different responsibilities between developed and developing countries and the US responsibility. The *common responsibility* discourse refers to China's responsibility, the responsibility of developing countries and the responsibility of major economies and emitters.

The 'Different Responsibilities' Discourse

Not surprisingly, one of fundamental China's positions on climate change is the responsibility of developed countries for addressing climate change. The advanced countries have been required to play a leading role in reducing emissions and providing financial and technological support to developing countries. Also, the developed countries have been blamed for their weak action on climate change. This fundamental position has prevailed over time in China. While China took a positive attitude towards addressing climate change particularly in 2015, it did not abandon its position on the historical, financial and technological responsibilities of developed countries.

Similarly, the position around the different responsibilities puts an emphasis on the responsibility of developed countries in addressing climate change issues. Compared to the position around the responsibilities of developed countries, it focuses on the principle of common but differentiated responsibilities between developed and developing countries and the key role of existing international climate change institutions such as the UNFCCC and the Kyoto Protocol.

China has emphasised that each country should adhere to the principle of common but differentiated responsibilities under the UNFCCC and the Kyoto Protocol. The principle was understood as a statement that developed countries have the historical responsibilities for reducing emissions and developing countries have the rights of economic development and poverty eradication. It is not only an important element of existing international climate change institutions but also it is one of fundamental positions of China. It has been emphasised mainly by governmental actors in the climate change discourse in China in 2015.

The US is recognised as having a key role in global climate change politics as it is a leading economy and greenhouse gases emitter. In the climate change discourse of China, a weak collective action on emission reduction is attributed to the lack of the US's role in global agreements particularly like the Kyoto Protocol. The US should be responsible for having a substantial action on reducing emissions. However, the climate cooperation between China and the US had been observed around the 2015 Paris climate conference. Discursively blaming the US had been transformed into a call for enhancing the bilateral cooperation.

The 'Common Responsibility' Discourse

China has received international pressures and has been criticised due to its fundamental principles of addressing climate change. In addition to the criticism, China has been required to have a proactive role in mitigating climate change. There is a discursive link between climate change and a responsible country. This demonstrates that with rapid economic growth, China had been seen as a major emitter, and it was called to undertake the responsibility for addressing climate change.

The developing countries have been discursively constructed to take the responsibility for addressing climate change. On the basis of the principle of common but differentiated responsibilities, developing countries, particularly emerging economies, should take the responsibility for addressing climate change as they increased the levels of carbon emissions dramatically. However, this is not to say that developed countries can shirk their responsibilities and obligations.

Emerging economies and major developing countries have been labelled as major emitters. They have thus been understood to take the collective action on addressing climate change. Firstly, emerging economies such as China and India were required to take action on substantial emission reduction as they are recognised as the major emitters. Secondly, China and the US were interpreted as two emitters to have responsibilities for addressing the climate issues. In this sense, China has been constructed to undertake the responsibility for addressing climate issues. In 2015, the close cooperation between the US and China had been observed. China and the US recognised their significant contributions to carbon emissions and their key roles in international climate change negotiations. However, this discourse raises an ambiguity of the differences between developed and developing countries.

Dynamic Discourse on the Responsibilities for Addressing Climate Change

The Dominance of Different Responsibilities Discourse

The discourse around the responsibility of developed countries and the different responsibilities had been identified in the climate discourse of China in 2007. Obviously, the discourse plays a dominant role in the climate governance of China. However, this is not to say that the common responsibility discourse did not exist in 2007. It had not yet been identified substantially in 2007. Therefore, the different responsibilities discourse played a dominant role in the discussions on climate change justice and responsibilities in China in 2007.

The Emerging Common Responsibility Discourse

While the *different responsibilities* discourse remained there, the *common responsibility* discourse had been emerging in 2009. It had been growing substantially in 2015. The responsibility of developing countries had been raised by China. In this sense, the discourse had been identified in 2015. But, it is worth noting that China does not abandon the principle of common but differentiated responsibilities. The principle remained a significant statement insisted by the governmental actors even in 2015.

Discursive Adoption in Climate Change Policy Rhetoric

The NDRC released China's National Climate Change Programme in June 2007. This was the first national policy to address climate change issues. It clarified the energy intensity target that China would reduce 20% energy consumption per unit GDP at the level of 2005 by 2010.

> To follow the principle of "common but differentiated responsibilities" of the UNFCCC. (NDRC, 2007, p. 24)

The document shows that the principle of 'common but differentiated responsibilities' had clearly reflected Chinese attitudes towards the responsibility for addressing climate change. This principle refers to a statement that while developed countries should take the responsibility of reducing emissions, developing countries have the rights of economic development and poverty eradication. This is a very fundamental principle of China in global climate politics.

The State Council announced China's target of cutting carbon intensity and its climate positions on 26th November 2009

> insisting the basic framework of the UNFCCC and the Kyoto Protocol with the principle of common but differentiated responsibilities (坚持《联合国气候变化框架公约》和《京都议定书》基本框架, 坚持"共同但有区别的责任"原则). (NDRC, 2009)

As the quotation demonstrates, China reiterated the principle of common but differentiated responsibilities and insisted the importance of the UNFCCC and the Kyoto Protocol recognised as the existing international climate institutions. This shows China's opposition to ideas for changing the framework of the UN in terms of global climate governance.

China submitted the proposal of INDCs to the institution of the UNFCCC on 30th June, 2015. China made a commitment to achieve the 2030 carbon peak target. In terms of the responsibility, China insisted on the importance of the principle of equity and common but differentiated responsibilities, and called the developed countries to play a leading role in addressing climate change. Also, China recognised developing countries as important actors to undertake the responsibility. It raised the importance of enhanced mitigation action and transparency in the document (NDRC, 2015, pp. 17–19).

Discussion and Conclusion

Discourse can affect political arrangements, provide policy options and interact with other discourses (Hajer & Wytske, 2013, p. 82; Hajer & Versteeg, 2005, p. 178). The domestic political discourse on the responsibilities can be socially constructed by a wide range of actors and has influenced China's substantial action on addressing climate change. China's shift in discourse on the responsibilities reflects its discursive adjustments in international climate change cooperation. Also, international pressures and expectations make a great contribution to the shape of the discourse on the different responsibilities. Especially since the 2008 global financial crisis, the role of China has been rising in international affairs and global governance such as the leadership in BRICS (Brizil, Russia, India, China and South Africa) and its important role in the APEC and G20.

In fact, there is nothing wrong with the categorisation of China into the non-Annex parties who have no obligations of compulsory emission reduction in the UNFCCC and the Kyoto Protocol. For example, the Group of Eight (G8), including the US, the UK, France, Germany, Canada, Japan, Italy and Russia which joined in 1998 and quitted after the Ukraine crisis in 2013, had played a leading role in global economic affairs and governance since 1975 (Hajnal, 2016, p. 33). Since 2005, climate change has become an important topic across the G8 summits

(Keohane & Victor, 2011, p. 11). Developed countries should their historical responsibilities for addressing the issues.

The discursive evolution from different responsibilities to common responsibility can be demonstrated by the close cooperation over climate change between the US and China. The US President Obama and Chinese President Xi vowed to take substantial climate action around the 2014 Summit on Asia–Pacific Economic Cooperation (APEC) (Dimitrov, 2016).

Future research should focus on how international discursive space has socially constructed the role of China and affects China's voice on global climate change negotiations. It is very interesting if China will have a great contribution to the development of global climate change governance and discourses.

References

Bang, G. (2015). The United States: Obama's push for climate policy change. In *The domestic politics of global climate change: Key actors in international climate cooperation*, 160.

Belis, D., Joffe, P., Kerremans, B., & Qi, Y. (2015). China, the United States and the European Union: Multiple bilateralism and prospects for a new climate change diplomacy. *Carbon & Climate Law Review*, 203.

Bjørkum, I. (2005). *China in the international politics of climate change: A foreign policy analysis*. The Fridtjof Nansen Institute.

Blaxekjær, L. Ø., & Nielsen, T. D. (2015). Mapping the narrative positions of new political groups under the UNFCCC. *Climate Policy, 15*, 751–766.

Brunnée, J., & Streck, C. (2013). The UNFCCC as a negotiation forum: Towards common but more differentiated responsibilities. *Climate Policy, 13*, 589–607.

Dimitrov, R. S. (2016). The Paris agreement on climate change: Behind closed doors. *Global Environmental Politics*.

Ferdinand, P., & Wang, J. (2013). China and the IMF: From mimicry towards pragmatic international institutional pluralism. *International Affairs, 89*, 895–910.

Foot, R., & Walter, A. (2010). *China, the United States, and global order*. Cambridge University Press.

Friman, M. (2013). Building legitimacy: Consensus and conflict over historic responsibility for climate change. In C. Methmann, D. Rothe, & B. Stephan (Eds.), *Interpretive approaches to global climate governance: (De)constructing the greenhouse*. Routledge.

Giddens, A. (2009). *The politics of climate change*.

Hajer, M., & Versteeg, W. (2005). A decade of discourse analysis of environmental politics: Achievements, challenges, perspectives. *Journal of Environmental Policy & Planning, 7*, 175–184.

Hajer, M., & Wytske, V. (2013). Voices of vulnerability: The reconfiguration of policy discourses. In J. S. Dryzek, B. N. Richard, & S. David (Eds.), *The Oxford handbook of climate change and society*. Oxford University Press.

Hajnal, P. I. (2016). *The G8 system and the G20: Evolution, role and documentation*. Taylor & Francis.

Harris, P. G. (2009). *World ethics and climate change: From international to global justice*. Edinburgh University Press.

Harris, P. G., Chow, A. S., & Karlsson, R. (2013). China and climate justice: Moving beyond statism. *International Environmental Agreements: Politics, Law and Economics, 13*, 291–305.

Harris, P. G., & Symons, J. (2013). Norm conflict in climate governance: Greenhouse gas accounting and the problem of consumption. *Global Environmental Politics, 13*, 9–29.

Heggelund, G. (2007). China's climate change policy: Domestic and international developments. *Asian Perspective, 31*, 191.

IPCC. (2014). Climate change 2014 synthesis report summary for policymakers. *The Fifth Assessment Report of the Intergovernmental Panel on Climate Change*. Intergovernmental Panel on Climate Change.

Keohane, R. O., & Victor, D. G. (2011). The regime complex for climate change. *Perspectives on Politics, 9*, 7–23.

NDRC, (2007). *China's national climate change programme*. National Development and Reform Commissions.

NDRC. (2009). *China announces targets on carbon dioxide emission cuts* [Online]. Department of Climate Change, National Developmentand Reform Commission. http://en.ccchina.gov.cn/Detail.aspx?newsId=38858&TId=123 [Accessed 23/01/2015].

NDRC. (2015). *Enhanced actions on climate change: China's intended nationally determined contributions*. NDRC.

Schreurs, M. A. (2016). The Paris climate agreement and the three largest emitters: China, the United States, and the European Union. *Politics and Governance, 4*, 219–223.

Stensdal, I. (2015). China: Every day is a winding road. In G. Bang, A. Underdal, & S. Andresen (Eds.), *The domestic politics of global climate change: Key actors in international climate cooperation*. Edward Elgar.

UN. (1992). *United Nations framework convention on climate change*.

WB. (2016). *Total greenhouse gas emissions*. The World Bank.

Winkler, H., & Rajamani, L. (2014). CBDR&RC in a regime applicable to all. *Climate Policy, 14*, 102–121.

WRI. (2014). *6 graphs explain the world's top 10 emitters* [Online]. World Resources Institute. http://www.wri.org/blog/2014/11/6-graphs-explain-world%E2%80%99s-top-10-emitters [Accessed 07/10/2016].

Wübbeke, J. (2013). China's climate change expert community—Principles, mechanisms and influence. *Journal of Contemporary China, 22*, 712–731.

Yu, H. Y., & Zhu, S. L. (2015). Toward Paris: China and climate change negotiations. *Advances in Climate Change Research, 6*, 56–66.

CHAPTER 5

Framing Climate Change Resilience in China

Abstract China is not only the largest greenhouse gases emitter, but it also is a victim of negative effects of climate change. Resilience to climate change has been adopted into China's media and policy discourse. This chapter reveals the main distinctive features of the coverage of climate resilience in China around the critical points of the International Panel on Climate Change (IPCC) reports released (2007, 2014 and 2018). The present study employs sectoral, moral and remedial frames to explore how the newspapers have constructed climate resilience issues, news sources and triggering events in a various way. The findings suggest that different newspapers have different features of constructing climate resilience in the coverage. Framing climate resilience in the coverage is routinely determined by the types of newspapers. Also, while the Chinese-written newspapers have decreased their attention to the issues, a national English newspaper has witnessed an increase in the coverage. An interesting finding reveals the term 'climate resilience' discursively exists very differently between Chinese and English texts. The study makes a primary contribution to media's framing climate resilience in China.

Keywords Climate resilience · Framing · Newspapers · Media · Discourse

S. Wang, *Climate Change Discourse in China*,
https://doi.org/10.1007/978-981-16-6754-1_5

Introduction

While China has been seen as an important role in global climate change politics being required to take a substantial action on reducing greenhouse gases emissions, it confronts the threats posed by the negative effects of climate change e.g. extreme weather events and rising sea level (Harris, 2011, pp. 8–13; Opitz-Stapleton, 2016). There is a wide range of existing literature on climate change communication and politics of China focusing on the dimension of mitigation (Bjørkum, 2005; Burgh & Zeng, 2010, pp. 17–19; Gallagher, 2007; Geall, 2018; Nisbet & Newman, 2015; Schreurs, 2010, pp. 96–97; Wübbeke, 2011). However, communicating climate resilience of China seems to have received weak attention from academic research.

The concepts of climate resilience and adaptation are not always interpreted in a same way. These two terms are not easy to distinguish from each other in the context of climate change issues because they have been used simultaneously and commonly. However, there is a nuance between adaptation and resilience but is a substantial implication for different pathways to actions on climate resilience. Resilience is linked more with a concern about climate disasters while it pays less attention to the individual (Wong-Parodi et al., 2015, pp. 5–6). Resilience could be understood as addressing large scale of natural disasters and disasters while adaptation could be seen as a smaller one (Wong-Parodi et al., 2015, p. 6). This chapter focuses on the concept of resilience researching how a complex and complicated system of climate resilience has been framed in the coverage of China.

In a nutshell, the concept of climate resilience involves two main dimensions. One is absorbing and resisting changes and returning to previous situations (Folke, 2006). Another dimension is transforming the social and economic structure for the changes into a different situation (Pelling, 2010; Walker et al., 2006). While facilitating climate resilience and transforming the economic and social structures plays a key role in addressing climate issues (ADB, 2012), this research does not elaborate the difference between the two dimensions of resilience. Rather, this chapter maps the climate resilience frames in the coverage of China.

The present study employs the frame approach developed by Entman (1993). To fit the approach with the requirement of this research, this chapter sets sectoral, moral and remedial frames as a pre-set coding scheme (see section on research methods). A majority of statements under each frame had been set in advance according to a deductive approach. However, in order to develop the coding scheme, this study also employs an inductive approach and creates new categorisations of statements.

Based on the literature review and discussion (see following section), this research develops a main question and several specific questions. The main research question is: What are the main features of the coverage of climate resilience in China? It looks how the coverage has been developing around the critical points of releasing the IPCC reports in 2007, 2014 and 2018.

Specific questions are set below:

- What type of newspapers have paid attention to climate change resilience?
- How have Chinese newspapers constructed climate change resilience through sectoral, moral and remedial frames?
- What are the differences in the coverage of the climate resilience between Chinese (*People's Daily*, *China Meteorological News* and *China Business News*) and English newspapers (*China Daily*) of China?
- What actors have been identified as news sources in the coverage of climate resilience in China?
- What are the associations between triggering events and the coverage of climate resilience in China?

Framing Climate Change Resilience

The concept of climate change resilience is not a straightforward and single interpretation of solutions to negative effects of climate threats. Instead, it involves a range of competing, conflicting and even ambiguous interpretations. Cannon and Müller-Mahn (2010) held a sceptical view of an emerging discourse on resilience. They preferred to the concept of vulnerability which focuses on people vulnerable to climate change and their social and economic justice. This is a moral consideration.

The concept of climate resilience is also interrogated by Boas and Rothe (2016) in terms of climate justice. In addition to vulnerability, the role of the developing countries is categorised as another moral consideration. In a Western perspective of climate security issues, the negative effects on poor nations could be seen as a concern about threats to the developed countries such as refugee crisis. This is so-called climate conflict discourse (Boas & Rothe, 2016). This view narrowly concentrates on the core national interests and security of the Western nations. It thus ignores the Western historical contributions to climate change and the inherent value of justice of providing assistance to the developing countries particularly those vulnerable to the natural disasters. This chapter categorises the concepts of vulnerability, the rights of developing countries and the responsibilities of developed countries into moral frame (see section on research methods).

However, it is important to note that the climate resilience discourse has changed adopting development and disaster management. The climate security concern discursively changed from an idea of defending within human conflicts to a concept of facilitating capability of addressing climate change in a resilient way (Corry, 2014). With the evolving interpretations of climate resilience, Aldunce et al. (2015) examined the resilience discourse within the framework of disaster management. They identify three storylines. First, technocratic storyline reflects an importance of information, knowledge and education. Second, community-based storyline includes participation, self-reliance of local governments and communities and social capital. Third, sustainability storyline refers to a concept of living with natural and environmental changes. However, while this categorisation of the resilience discourse raises what they need to do for addressing climate resilience issues, it requires a further discussion on how they can do.

As Methmann and Oels (2015, p. 54) summarise, three types of resilience are identified. First, engineering resilience refers to returning to previous situations. This emphasises human capability of utilising technologies for confronting the negative effects of climate change. Second, ecological resilience refers to a concept of requiring local communities to manage the climate disaster in a long term. Third, socio-ecological resilience emphasises the importance of transforming political, economic and social structure and systems in order to confront climate change issues. Being different to the ideas raised by Aldunce et al. (2015), it seems the third concept of resilience requires a comprehensive change in social and political structure rather than solely emphasising the importance of technology and local participation.

However, these categorisations of the climate resilience fail to take development issues and the rights of developing countries into consideration. Cannon and Müller-Mahn (2010) criticise the climate resilience discourse for ignoring the basic interests and needs of people in developing countries. They explicitly call for raising attention to links between development and climate resilience. Climate resilience action should be rooted in those who are vulnerable to the effects of climate change. Saigal (2014, p. 21) attaches an importance to links between poverty eradication and climate resilience. Climate resilience programmes and criteria not only include climate disaster management and ecosystem but also contain poverty, development and food security. Therefore, climate resilience discourse is required to adopt development issues and concerns. In other words, development is identified as a means of addressing climate resilience. Based on the existing literature, this chapter explores what kinds of issues, fields and sectors have been linked to climate resilience identifying sectoral frame. And, it also categorises the solutions into remedial frame (see section on research methods).

While mitigating climate change requires a global collective action, resilience is substantially implemented at local level. In this sense, the objective referent of constructing the concept of climate resilience shifted from sovereign states to various social stakeholders (Boas & Rothe, 2016). Various stakeholders have participated in constructing climate resilience discourse including governmental actors, economic policy-makers, non-government organisations (NGOs) and academic communities (Methmann & Oels, 2015, p. 629). Gidley et al., (2009, p. 2) categorised the definitions of climate resilience into two dimensions. One is translating a passive cope into resilience action. This emphasises that local people and stakeholders passively receive information but lacks a proactive role in understanding climate resilience knowledge. Another one is introducing mutual learning and conceptual development to active resilience emphasising the importance of dynamic associations between subjective meanings and objective effects. With the involvement of various stakeholders in addressing climate change resilience, they should be encouraged to proactively construct the resilience discourse.

However, Fisichelli et al., (2016, p. 754) state that the term of climate resilience can be interpreted by various social actors in a very different way. They review the various interpretations of climate change resilience and identifies three ways of understanding resilience namely curbing change and returning to previous situations, adjusting to change and transforming social and economic structure (Carpenter et al., 2001; IPCC, 2007; Fisichelli et al., 2016, p. 755). While the various interpretations have an inherent value of providing solutions and expressing their distinctive interests and concerns, they obviously lead to competing, conflicting and complicated discourses and ideas placing barriers to climate resilience (Eisenack et al., 2014). Therefore, this research looks at how different actors have been cited as news sources constructing their various ideas in the coverage.

Climate Resilience Journalism

As discussed above, the various interpretations made by various social stakeholders lead to competing, conflicting and complicated ideas of implementing climate resilience (Anderson, 2014). This situation might cause a difficulty in effectively communicating knowledge and information of climate resilience.

Although a substantial global action on mitigating carbon emissions remains far sufficient to avoid a catastrophic consequence, it has received unprecedented attention from policy-makers, business, academic communities, pressures groups and particularly media. However, as Moser (2017) explains, these stakeholders show little suggestions of improving the climate resilience communication mainly because substantial adaptation and resilience policies just have emerged very recently.

Therefore, this chapter shows a dynamic process of climate resilience discourse constructed by the media. There are three features and distinctive contributions of this research. First, it is innovative to discover how the various interpretations have been framed in the news communication. Second, it is important to identify the various social stakeholders in the coverage of climate resilience. Third, this chapter observes how climate resilience have discursively changed in the communication and coverage. In order to respond to the three points, this chapter employs the newspapers as the data source of media. The newspapers can show how various stakeholders have been cited as news sources (Carvalho, 2008), and frames have emerged through journalistic practices (Bohensky & Leitch, 2014; Boykoff & Boykoff, 2007). The method section raises specific justifications of using newspapers and identifying climate resilience frames and discourse.

Climate Resilience in China

China has witnessed its increase in carbon emissions and is identified as the largest emitter. While China, being a developing country, does not have historical responsibilities for addressing climate change issues, its per-capita emissions are even nearly reaching the level of the EU (Torney, 2015, p. 110). China is a key global economy and is required to undertake responsibility for global climate change governance (Ellermann & Mayer, 2009).

While mitigation actions focus on improving domestic energy and economic structures, climate resilience can be understood as global affairs. The vulnerable nations are the victims of effects of climate change while they do not have financial and technological capabilities to address the negative effects. These countries have been raising attention from international society calling for helps and supports. Being a developing country, China shows its positions on defending the basic rights of these countries (Schreurs, 2011).

In the national context, China itself is a victim of climate change issues recognising the associations between climate change and human security and development. Also, China identifies itself as a nation vulnerable to climate change and raises concerns about economic, energy, ecological, food and even human security (Opitz-Stapleton, 2016).

In the political context, China has raised the importance of climate resilience on its political agenda. In 2007, China released *the National Climate Change Programme* emphasising the importance of addressing climate issues. In 2013, China released *the National Strategy for Climate Change Adaptation*. In 2014, China released *the National Plan on Climate Change (2014–2020)*. Addressing the effects of climate change has explicitly been understood as an important element of national policies (Kostka & Zhang, 2018).

Based on the above discussions, there is an academic gap between framing climate resilience in the coverage and the case of China. The main objective of this chapter is to map the various frames identified in the coverage of China, observe the features of different newspapers constructing climate resilience issues, and discover how various social actors and triggering events have been constructed in the coverage.

Research Methods

Identifying Frames

Entman (1993) employs the concept of frame to categorise four types of statements namely problems, causes, moral judgements and remedies/solutions. While the categorisation has been widely applied to the studies on climate politics and commutation (Nisbet, 2009), this work raises an innovative approach to the employment of categorising frames. This is mainly because climate resilience itself is a concept of reacting to the problems and effects of climate change. Thus, this chapter discovers how specific sectors and/or fields have been related to climate resilience. Also, the coverage of climate resilience does not focus on discussing the causes while it recognises the reality of human-related climate change and raises the importance of action on addressing the issues. Therefore, this chapter identifies three types of frames namely sectors/fields, justice/moral judgements and remedies/solutions (see Table 5.1).

Sectoral frame refers to the statements discussing climate resilience from various functional and sectoral perspectives. For example, addressing the negative effects of climate change can be conducted within the agricultural sector when farmlands and agricultural production are threatened by extreme weather events or water shortage.

Moral frame refers to the statements of identifying the obligations of and responsibilities for addressing climate resilience. This research identifies the statements 'vulnerability' and 'developing countries'. The statement 'vulnerability' refers to an emphasis on the people and places vulnerable to effects of climate change. The statement 'developing countries' refers to the responsibilities of developed countries for providing supports to developing countries and the importance of the developing countries to prioritise economic development and poverty eradication.

Remedial frame refers to the statements of raising solutions to resilience to climate change. There are various interpretations of the remedial frame indicating different solutions. For example, the statement 'infrastructure' refers to strengthening the capabilities of resilience to climate change through infrastructure construction. However, the statement 'disaster management' emphasises reactions to the outcomes of natural disasters caused by climate change. It is important to raise the case of the statement 'adaptation'. This statement is identified in the texts of the coverage but it does not show detailed information and specific solutions. This statement can be identified particularly at the early stage

Table 5.1 Climate change resilience frames and statements

Coding term	*Statements*
Sectoral frame	
Agriculture	Agriculture is closely related to climate resilience
Cities	Cities are closely related to climate resilience
Community	Local communities are closely related to climate resilience
Costal	Coastal issues threatened by rising sea level should be addressed
Ecosystem	Conserving ecosystem is a key to climate resilience
Forest	Forest is a key to climate resilience
Industrial sectors	Climate resilience is implemented across various industrial sectors
Public health	Public health is closely related to climate resilience
Regional	Climate resilience is translated into regional strategies
Tourism	Tourism is closed is closely related to climate resilience
Water	Water resources, rivers and lakes are closely related to climate resilience
Remedial frame	
Adaptation	The 'adaptation' is just mentioned in the texts and thus lacks of a clear explanation
Development	Climate resilience takes social and economic development into consideration
Disaster management	Climate resilience relies on disaster prevention and reduction management
Infrastructure	Strengthening infrastructure is a key to climate resilience
Security	Climate resilience is an urgent issue to address
Moral frame	
Developing countries	Developing countries have rights of development and poverty eradication. Developing countries should receive supports from other countries
Vulnerability	People or places are vulnerable to effects of climate change

of the coverage of climate resilience. It is important to identify the remedial frame because different interpretations and ideas can imply various solutions and even affect public understandings of climate resilience.

This research employs inductive and deductive approaches to identifying the statements and categorising the frames. On the one hand, the deductive approach to coding data is used by developing a pre-set coding schemes based on the findings of the existing literature review. These coding terms include 'vulnerability', 'security', 'disaster management', 'community' and 'development' (see section on result). On the other hand, this research employs an inductive approach to identifying new statements and frames during the coding process. This chapter pays attention to how sectors and fields have been linked to climate resilience in the coverage of China.

Data Collection

The present study identifies the climate-related news articles through searching the database of National Knowledge Infrastructure (CNKI). The CNKI provides the data of Chinese newspapers, academic articles and journals. The data from *China Daily* is collected through its official online database due to its absence in the CNKI. It is important to note that the number of news articles across the newspapers (published by November, 2018) collected in the CNKI through inputting the key term 'climate change adaptation' (qihou bianhua shiying) is very small (N = 32). As a result of a further attempt with inputting the key terms 'climate change AND adaptation', there is 198 news articles. Therefore, this research firstly collected the data with the key term 'climate change' (qihou bianhua) and secondly identify the term 'adaptation' (shiying) mentioned and clarified in the texts of the newspapers. As a result, the top ten newspapers of most frequently reporting climate change issues can be identified (see section on types of newspapers). The data collection does not exclude comments and opinions. This is because climate resilience relates to professional knowledge and scientific terms which can be identified within the professional comments of the coverage. Also, while the Chinese characters 'renxing' ('resilience') cannot be identified in the Chinese context of climate change issues, this work collects data through inputting 'resilient cities' which has been frequently used in the coverage in China (see section on 'resilient cities').

This work selected the four representative newspapers of China in terms of the coverage of climate resilience. Due to the reform in the media system of China, different newspapers receive different levels of state control and marketisation (Stockmann & Gallagher, 2011; Zhao, 2004). The three papers are Chinese-written ones selected from the top ten, and one English-written paper is an influential and nationally circulated paper in China. They are *People's Daily*, *China Meteorological News*, *China Business News* and *China Daily*. *People's Daily* is the most important newspaper in China as it is the mouthpiece of the Chinese government and the Communist Party of China (Lv, 2014). *China Meteorological*

News is a functional newspaper focusing on the coverage of meteorological issues, weather events and scientific research on climate change. It is operated by the China Meteorological Administration (CMA) and provides information of climate-related news. *China Business News* is identified as the source of data collection because it has two distinctive features. First, this newspaper focuses exclusively on economic and business news, and it can be used to find how the climate resilience has been framed within economic field. Second, it is circulated nationally but its base is in Shanghai which is the financial centre of China. While *China Economic Herald* is based in Beijing and is supervised by the National Development and Reform Commission (NDRC) routinely reflecting governmental positions, *China Business News* is a marketised newspaper and can reflect various voices. *China Daily* is an English-written newspaper circulated in China. While *China Daily* represents the governmental position of China, it is a marketised newspaper having a wider space of reporting news. Therefore, these four different newspapers are selected as the sources of data collection in terms of the coverage of climate resilience.

This work raises three critical points of addressing climate change issues. In contrast to mitigation focusing on the causes of climate change, adaptation/resilience can be directly linked to the negative effects of the issues. The IPCC reports have an explicit effect on climate communication and frames (O'Neill et al., 2015). Therefore, this work identifies the years of releasing the IPCC reports as the critical points to collect data. The IPCC had released its fourth report in 2007, its fifth synthesis report in 2014 and a special report on global temperature rise to 1.5°C in 2018 (IPCC, 2018).

This work collects data from the coverage of *China Daily* published in 2007, 2014 and 2018 respectively. As a result, the number of the identified and selected news articles is 9 in 2007, 45 in 2014 and 47 in 2018. *China Meteorological News* shows 119 news articles totally while this work finally identifies and selects 8 articles in 2007, 7 in 2014 and 8 between 2017 and 2018. The year 2017 is considered simply because this newspaper does not show a substantial attention to this topic in 2018. Due to a small number of the news articles related to climate resilience identified in the coverage of *People's Daily* and *China Business News*, this work identifies and collects the data of the two papers published from 2007 to 2018. As a result of the data collection, there are 41 news articles identified in

People's Daily and only 10 in *China Business News*. For these four newspapers, the latest date of the news reports identified is 27th November, 2018 when the work of data collection ended. However, this would not affect the analysis and expectations of the research while it focused mainly on a qualitative approach to identifying frames.

Results

Types of Newspapers

The top ten newspapers of frequency of reporting climate change issues are circulated nationally (see Table 5.2). This demonstrates that climate change issues have mainly been reported at the national level in China. Among them, there are two significant newspapers of China namely *People's Daily* and *Xinhua Daily Telegraph* supervised by Xinhua News Agency. There are five functional newspapers which are supervised by specific ministerial agencies namely *China Meteorological News, Science and Technology Daily, China Environmental News, China Energy News* and *China Green times*. For example, *China Meteorological News* is supervised by the CMA (see section on data collection). This indicates that climate change discussions have mainly been reported by the functional newspapers. There are two economy-related newspapers namely *China Economic Herald* and *China Business News*. It shows that climate change issues have been linked to economic and financial topics in the coverage. At the bottom of the list, it is *Chinese Social Sciences Today* which is an

Table 5.2 The top ten Chinese newspapers of frequently reporting climate change issues

Newspapers	*Type*
China Meteorological News (中国气象报)	Functional
People's Daily (人民日报)	Central
Science and Technology Daily (科技日报)	Functional
China Environmental News (中国环境报)	Functional
China Economic Herald (中国经济导报)	Economic
Xinhua Daily Telegraph (新华每日电讯)	Central
China Business News (第一财经日报)	Economic
China Energy News (中国能源报)	Functional
China Green times (中国绿色时报)	Functional
Chinese Social Sciences Today (中国社会科学院报)	Academic

academic newspaper. It demonstrates that the climate issues have been discussed by social scientists in China.

Figure 5.1 shows the number of articles identified and filtered from the initial data collected. The articles not relevant to the research are filtered and excluded. For example, while one news article mentions the term 'economic adaptation' and raises an emphasis on adjusting economic system, it is not relevant to the topic on climate change issues. *People's Daily* showed its gradual decline in the coverage of climate resilience. *China Meteorological News* and *China Business News* keep stable but quite weak attention to the topic on climate resilience. However, a dramatic increase in the coverage is identified in *China Daily*, even though its decline occurred in 2018. The decline observed can partly be explained by the set of data collection. The date of finishing the collection was before the 2018 climate change conference which might trigger a number of climate-related news reports (see section on data collection). In summary, the trend in the coverage of climate resilience in China shows a gap between Chinese-written and English-written newspapers.

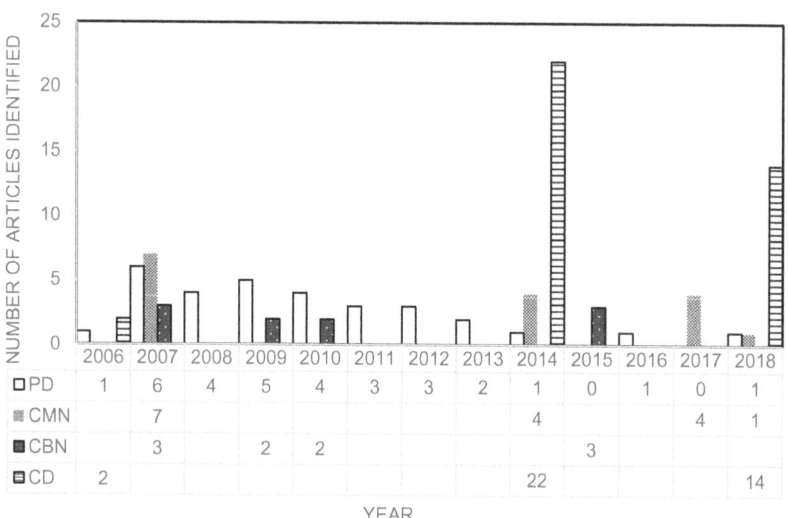

	2006	2007	2008	2009	2010	2011	2012	2013	2014	2015	2016	2017	2018
PD	1	6	4	5	4	3	3	2	1	0	1	0	1
CMN		7							4			4	1
CBN		3		2	2					3			
CD	2								22				14

Fig. 5.1 The coverage of climate change resilience in the four newspapers (*Note* PD denotes *People's Daily*; CMN denotes *China Meteorological News*; CBN denotes *China Business News*; CD denotes *China Daily*)

Framing Climate Change Resilience in People's Daily, China Meteorological News and China Business News

People's Daily

People's Daily has framed climate change resilience into various concerns and sectors over time (see Appendix A). While *People's Daily* has witnessed its decline in the coverage of climate resilience, its focus has shifted from simply mentioning the term 'adaptation' to outlining the specific sectoral frame such as 'cities' and the remedial frame such as 'infrastructure' and 'disaster management'. Particularly, the Ministry of Housing and Urban–Rural Development was cited as a news source emphasising the importance of enhancing capabilities of infrastructure resilient to climate change.

Being the most important paper of China, not surprisingly, *People's Daily* links climate resilience to the moral frames 'developing countries' and 'vulnerability'. On the one hand, the paper emphasises that China and other developing countries are the victims of climate change, and they are vulnerable to the negative effects. On the other hand, it focuses on the responsibilities of developed countries for providing financial and technological supports to the vulnerable nations in terms of climate change issues.

China Meteorological News

China Meteorological News is a functional newspaper supervised and managed by the CMA. The majority of statements are identified to be around the sectoral and remedial frames (see Appendix B). For example, the statement 'disaster management' can be identified in the paper frequently. Other statements like 'infrastructure', 'water' and 'ecosystem' have been widely linked with the news sources from ministerial agencies, academic institutions and local governmental agencies. This demonstrates that *China Meteorological News* substantially focuses on how climate change resilience can be addressed and implemented.

China Business News

While *China Business News* is widely engaged in financial, economic and business topics, it does not link climate resilience to specific and practical solutions. Rather, the paper overwhelmingly raises climate resilience as the moral frame (see Appendix C).

> Developed countries need to provide 200 billion dollar every year in order to help poor nations adapt to climate change (发达国家每年需要拿出超过2000亿美元, 以帮助贫穷国家适应气候变化). (see 07/12/2009, *China Business News*)

The above statement shows that the paper linked the climate resilience of developing countries to the ambitious commitments to the financial supports from developed countries. A debate on climate justice has been made around the responsibilities and obligations of the developed countries.

Framing Climate Resilience in China Daily

In contrast to the papers written in Chinese having a decline in the coverage of climate resilience, *China Daily*, written in English, has paid attention to the topics even in 2014 and in 2018 (see Appendix D). There are three developments of framing climate resilience identified in the coverage of *China Daily*.

First, the terms 'adaptation' and 'resilience' can be identified in the coverage of *China Daily* clearly. The term 'resilience' has been increasingly used in the coverage of *China Daily* (see Fig. 5.2). It has been identified within sectoral and remedial frames. For example, 'climate-resilient development' refers to a concept of combining economic and social development and climate resilience. An increase in the frequency of using the term 'resilience' reflects proactive and specific solutions of climate resilience constructed in the coverage. As Boykoff et al. (2013) discussed, existing studies focus on the explicit mentions of adaptation to climate change in the mass media. However, this research observed a decline in using the term 'adaption' which has been simply mentioned in the coverage.

Second, the remedial statement 'development' has been emerging in the coverage.

> The idea of a "smart city" is to create intelligent and efficient solutions to support growth toward a more sustainable and resilient society. (see 'Forum gets smart with city issues', 22/02/2018, *China Daily*)

The above statement was found in the coverage of *China Daily* establishing a link between economic growth and climate-resilient society.

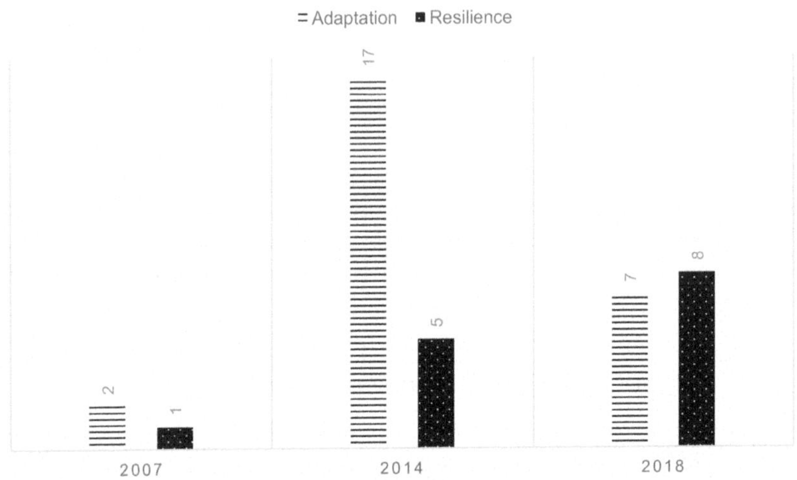

Fig. 5.2 The number of articles with the terms 'adaptation' and 'resilience' identified in the coverage of *China Daily*

Framing the concept of development into climate resilience can make a potential for encouraging social stakeholders to support actions on addressing the negative effects of climate change. For example, the stakeholders can have incentives to be engaged in constructing the climate-resilient society in accordance with the concept of sustainable development.

Third, the moral frame has obviously declined in the coverage in 2018. This indicates a proactive attitude towards climate resilience identified in *China Daily*. A heavy emphasis on justice of climate resilience reflects a fierce debate on the responsibilities and obligations rather than a talk about specific and practical solutions and ideas.

'Resilient Cities'

While the term 'resilience' cannot be straightforwardly identified in Chinese in terms of climate change issues, it has been reflected within the concept of resilient cities in the Chinese-written newspapers. There are two ways of describing the resilient cities in Chinese namely 'renxing chengshi (韧性城市)' or 'tanxing chengshi (弹性城市)'. The number of articles with 'renxing chengshi' is totally 33 and that with 'tanxing chengshi' is 47 identified across the newspapers in the CNKI. The number of articles selected and filtered is 21 and 17, respectively. Strictly speaking, the characters 'renxing chengshi' seem to effectively reflect the nature of the resilient cities. This is because the characters are the outcome of directly translating the term from English to Chinese. While the characters 'tanxing chengshi' sound like 'a flexible city' in Chinese, they are widely used as the idea of resilient cities.

The earliest coverage with the concept of resilient cities was identified in 2011 when the first Mayors Summit on Disaster Prevention and Reduction was held in Chengdu, China. The concept was not only linked to the disaster management but it also was framed as a solution to climate change resilience. A majority of the coverage related to resilient cities had been identified since 2017. This demonstrates that the concept of resilient cities has been adopted and constructed in the coverage of China very recently.

Types of Actors/Sources Cited in the Coverage

Various actors have been cited as news sources in the coverage of climate resilience of China (see Appendices A, B, C and D). The coverage of *People's Daily* substantially relies on governmental actors, foreign governments and international governmental organisations as main news sources. *China Meteorological News* focuses on professional and scientific news related to meteorological issues. It thus cites CMA and academic institutions as news sources talking about technocratic solutions to climate resilience. Being different to the mouthpiece and the functional paper, *China Business News* has a wide space for having contact with various social stakeholders. The main news sources of the coverage are governmental actors, international organisations, foreign governments and NGOs. *China Daily* has a much wider space for reporting news. The majority of the news sources cited in the coverage of climate resilience are international organisations, foreign governmental actors, NGOs and academic institutions.

The majority of news sources cited in the coverage of 'resilient cities' are international organisations, academic institutions and local actors of China. The international organisations, like the United Nations Human Settlements Programme, emphasising the importance of resistance to external shock and fast restoration of cities. Academic institutions focus on research and technocratic solutions to climate resilience. Local actors emphasise how cities can be constructed to be resilient to the natural disasters such as flooding. Also, they discuss how they integrate the concept of resilient cities into local development strategies.

Associations Between Triggering Events and Climate Resilience Coverage

Different newspapers show different features of the associations between triggering events and climate resilience coverage (see Appendices A, B, C and D). *People's Daily* reflects the national actions and policies and international climate change conferences. The coverage of *People's Daily* had been substantially triggered by the events of releasing governmental reports and policies in terms of addressing climate change.

China Meteorological News shows its coverage trigged by the events of releasing the policies and reports of climate change. However, the coverage of this functional paper related to meteorological affairs had not been triggered by the events of extreme weather and natural disasters caused by climate change. This requires a further study on why climate change resilience have not been linked to the weather and disaster events.

The coverage of *China Business News* had been triggered mainly by the UN climate change conferences, reports and talks. This is mainly because this paper has concentrated on topics on financial and technological supports to the developing countries. These topics are main arguments and debates during the periods of international climate conferences and negotiations.

This research finds a substantial change in the coverage of *China Daily* in terms of the associations. In 2007, the triggering events were identified as a business event and a governmental action. In 2014, the coverage had substantially triggered by the international events particularly the UN climate conferences and the release of the fifth IPCC report. In 2018, the coverage was triggered by specific events. The events around the topics on rivers, ocean and smart cities had been identified in the coverage. This demonstrates *China Daily* has shifted its attention from national policies and international politics to specific, practical and local solutions.

Not surprisingly, the coverage of 'resilient cities' has been closely linked to the cities and infrastructure, and its coverage had been triggered by the conferences and forums on cities and urban construction.

Discussion and Conclusions

The functional newspapers play a dominant role in constructing climate resilience in China. This demonstrates the climate resilience issues remain the topic raised by the newspapers (*China Meteorological News*; *Science and Technology Daily*; *China Environmental News*) who are supervised by the functional governmental agencies related to climate resilience. In addition to the functional newspapers, *People's Daily* and *Xinhua Daily Telegraph*, the central newspapers, pay attention to the issues (see Table 5.2). This demonstrates the climate resilience remains constructed at the national level while the issues physically occur at the sub-national level. However, compared to emissions reductions, climate resilience requires more substantial attention and actions at local levels.

There is a decrease in the coverage of climate resilience identified in the Chinese-written newspapers. It seems that climate resilience has been losing attention from the newspapers in China. However, indeed, this low level of media's attention to the issues can be explained by their focus on other climate change issues e.g. carbon emissions, responsibilities for reducing emissions and low carbon economy. As Moser (2017) states, a lack of studies on climate resilience communication is mainly because the resilience policies have not been emerging until recently. China released *the National Strategy for Climate Change Adaptation* in 2014. This requires a further research on how the developments of policy discourse impacts on the coverage of climate resilience in China.

People's Daily shows its consistency in framing climate resilience into specific solutions particularly like disaster prevention and reduction management. This reflects a close link between climate change resilience and natural disasters. A main reason for this framing can be explained by the finding that while the CMA does not play a key role in the Chinese governmental system, it has been frequently and routinely cited as the main news source in the coverage. The CMA has capability of providing professional information and concepts of climate resilience. On the other hand, *People's Daily* defends the fundamental position of Chinas and raises its concerns about the rights and interests of developing countries.

Being contrast to *People's Daily*, *China Meteorological News* pays less attention to justice. First, the paper does not cite a wide range of actors engaged in international climate change negotiations such as the Ministry of Foreign Affairs (MoFA) as news sources. Even though the ministerial agencies, like NDRC and MoST, are identified in the coverage, they were cited to embrace the remedial frames such as disaster management. Second, the term 'climate-resilient city' (qihou shiyingxing chengshi) can be identified in the coverage demonstrating the implementation of the climate resilience at local administrative levels in China. While the paper pays attention to local climate resilience citing local governments as news sources, it does not have much space for raising the moral frame constructed at the international level.

It is an interesting finding that *China Business News* substantially constructs the importance of the responsibility of developed countries for providing financial support to vulnerable nations in terms of climate adaptation and resilience. The coverage had paid considerable attention to the 2009 Copenhagen Climate Conference and the 2015 Paris Conference with a focus on commitments to climate financing. This is simply because the newspaper focuses on the financial issues of climate change which have been debated around the moral considerations in climate change negotiations. Another explanation of the feature of the coverage is citing the NGOs like Oxfam and The Climate Group as news sources. The international environmental NGOs can be very effective to attract media's attention through various activities and raise their claims of environmental justice (Cox & Schwarze, 2015).

This chapter offers a distinctive finding in the comparison between the Chinese and English newspapers in China. In the papers written in Chinese, the difference between these two terms cannot be identified while they are expressed in the same term 'adaptation' (shi ying). Even though the term 'climate-resilient cities' can be identified in the Chinese-written media, their Chinese characters indeed mean 'climate-adaptive cities' (qihou shiyingxing chengshi). In other words, there is more or less inconsistent in expressing the term 'resilience' between Chinese characters and English words.

In the coverage of *China Daily*, climate change resilience has been integrated into the consideration of economic and social development. Resilience requires society to transform into a climate-resilient situation. Also, *China Daily* shows that various sectoral issues such as 'water' and 'public health' and remedial solutions such as 'infrastructure' had been framed in the coverage while the decline of the moral frames was identified in 2018. This demonstrates a shift from moral discussions to specific solutions constructed in the newspapers. Various international actors are identified as the news sources in the coverage. This is because this paper is written in English having an open attitude towards international affairs and news sources.

There is not a distinctive difference between 'renxing chengshi' and 'tanxing chengshi' in terms of climate-resilient cities. However, there are two nuance differences. First, the concept of sponge city, which is frequently used as a solution to extreme weather events and natural disasters, is substantially framed into the categorisation of 'tanxing chengshi'. Second, the concept of 'renxing chengshi' has been widely used as a situation of resisting to external shocks, restoring and transforming into a different place under the threat of climate change. The concept of 'tanxing chengshi' has been identified as a situation of absorbing disturbance and returning to its previous condition. While the two concepts refer to the term 'resilient cities', they imply very different solutions and strategies. The difference reflects a lack of dialogues between climate change and urban construction.

In terms of the associations between the triggering events and the coverage of climate resilience, three findings can be summarised here. First of all, the coverage of climate resilience in China is not triggered by the events of extreme weather and natural disasters. This demonstrates that the newspapers of China have not constructed a close association between the extreme weather events and the importance of climate resilience. This might cause a lack of public awareness of the importance of climate resilience. Second, international conferences and national reports and policies are the main triggering events of the coverage in China. This indicates that the newspapers have not identified and constructed the associations between climate resilience and sub-national events. This is not a positive signal for the climate resilience required to be implemented at the local levels. Third, it is important to note that the coverage of resilient cities is closely associated with the events of specific and professional discussions. The associations between climate resilience and urban construction had received attention from the newspapers of China. This implies that the effective communication of climate resilience requires a close link between the concept of resilience and specific industrial sectors.

In conclusion, while Chinese-written newspapers, *People's Daily*, *China Meteorological News* and *China Business News,* have decreased their attention to the issues, a national English newspaper, *China Daily*, has an increase in the coverage. The types of the newspapers determine their different ways of framing climate resilience. *People Daily* is the mouthpiece of China constructing climate resilience across sectoral, remedial and moral frames. It not only concentrates on identifying problems and solutions but also raising the differentiated responsibilities between developed and developing countries. *China Meteorological News* is a functional paper of reporting meteorological affairs. In this sense, it focuses on constructing problems and solutions of climate resilience but pays less attention to discussions on climate justice. *China Business News* is a marketised newspaper focusing on financial and economic news. It thus concentrates on the financial dimension of climate resilience which has been debated substantially around the moral discussions. *China Daily* is a national paper written in English, and it thus has a strong link to international affairs (Yi-jun et al., 2011). While climate resilience has been increasingly discussed at global level, its rise can be identified in the coverage of *China Daily*. Various actors have been identified across the newspapers as the news sources in the coverage.

This research requires a further study on a long-term observation of the communication of climate resilience in China. It is very important to find out how the issues can be framed in different newspapers and how different actors have been cited as news sources (Anderson, 2015). Also, the interesting finding reminds academic communities and policy-makers of raising a creative perspective of developing the concept of climate resilience in Chinese. The research identifies a clear gap between English and Chinese texts in terms of the concept. More importantly, a further research will need to discover why the coverage of climate resilience has a very weak link to the natural disasters and extreme weather events (Nature, 1934).

References

ADB. (2012). From adaptation to resilience: The need for transformational change—Bindu N. Lohani. *Asian Development Bank*.

Aldunce, P., Beilin, R., Howden, M., & Handmer, J. (2015). Resilience for disaster risk management in a changing climate: Practitioners' frames and practices. *Global Environmental Change, 30*, 1–11.

Anderson, A. (2014). *Media, environment and the network society*. Springer.

Anderson, A. (2015). Reflections on environmental communication and the challenges of a new research agenda. *Environmental Communication, 9*, 379–383.

Bjørkum, I. (2005). *China in the international politics of climate change: A foreign policy analysis*. The Fridtjof Nansen Institute.

Boas, I., & Rothe, D. (2016). From conflict to resilience? Explaining recent changes in climate security discourse and practice. *Environmental Politics, 25*, 613–632.

Bohensky, E., & Leitch, A. (2014). Framing the flood: A media analysis of themes of resilience in the 2011 Brisbane flood. *Regional Environmental Change, 14*, 475–488.

Boykoff, M., Ghosh, A., & Venkateswaran, K. (2013). Media discourse on adapation: Competing vision of "success" in the Indian context. *Successful Adaptation to Climate Change.*

Boykoff, M. T., & Boykoff, J. M. (2007). Climate change and journalistic norms: A case-study of US mass-media coverage. *Geoforum, 38*, 1190–1204.

Burgh, H. D. & Zeng, R. (2010). *New opportunities for environmental and climate change journalism in China*. International Media Support.

Cannon, T., & Müller-Mahn, D. (2010). Vulnerability, resilience and development discourses in context of climate change. *Natural Hazards, 55*, 621–635.

Carpenter, S., Walker, B., Anderies, J. M., & Abel, N. (2001). From metaphor to measurement: Resilience of what to what? *Ecosystems, 4*, 765–781.

Carvalho, A. (2008). *Communicating climate change: Discourses, mediations and perceptions.*

Corry, O. (2014). From defense to resilience: Environmental security beyond neo-liberalism. *International Political Sociology, 8*, 256–274.

Cox, R., & Schwarze, S. (2015). Strategies of environmental pressure groups and NGOs. In A. Hansen (Ed.), *The Routledge handbook of environment and communication*. Routledge.

Eisenack, K., Moser, S. C., Hoffmann, E., Klein, R. J. T., Oberlack, C., Pechan, A., Rotter, M., & Termeer, C. J. A. M. (2014). Explaining and overcoming barriers to climate change adaptation. *Nature Climate Change, 4*, 867–872.

Ellermann, C., & Mayer, M. (2009). *Climate change with Chinese characteristics: A study of discourse*. IOP Publishing, 572025.

Entman, R. M. (1993). Framing: Toward clarification of a fractured paradigm. *Journal of Communication, 43*, 51–58.

Fisichelli, N. A., Schuurman, G. W., & Hoffman, C. H. (2016). Is 'resilience' maladaptive? Towards an accurate Lexicon for climate change adaptation. *Environmental Management, 57*, 753–758.

Folke, C. (2006). Resilience: The emergence of a perspective for social–ecological systems analyses. *Global Environmental Change, 16*, 253–267.

Gallagher, K. S. (2007). China needs help with climate change. *Current History, 107*, 389.

Geall, S. (2018). Climate-change journalism and "edgeball" politics in contemporary China. *Society & Natural Resources, 31*, 541–555.

Gidley, J., Fien, J., Smith, J.-A., Thomsen, D., & Smith, T. (2009). Participatory futures methods: Towards adaptability and resilience in climate-vulnerable communities. *Environmental Policy and Governance, 19*, 427–440.

Harris, P. G. (2011). Diplomacy, responsibility and China's climate change policy. In *China's Responsibility for Climate Change: Ethics, Fairness and Environmental Policy*.

IPCC. (2007). *Climate change 2007: Synthesis report*.

IPCC. (2018). *Global warming of 1.5 °C*.

Kostka, G., & Zhang, C. (2018). Tightening the grip: Environmental governance under Xi Jinping. *Environmental Politics, 27*, 769–781.

Lv, S. (2014). *Chinese newspaper: Transformation under the view of market and internet (Zhongguo Baoye: Shichang Yu Hulianwang Shiyuxiade Zhuanxing)*. Social Science Academic Press (China).

Methmann, C., & Oels, A. (2015). From 'fearing' to 'empowering' climate refugees: Governing climate-induced migration in the name of resilience. *Security Dialogue, 46*, 51–68.

Moser, S. C. (2017). *Communicating climate change adaptation and resilience*. Oxford University Press.

Nature,. (1934). Will the Chinese mitten crab invade British waters? *Nature, 134*, 17–17.

Nisbet, M. C. (2009). Communicating climate change: Why frames matter for public engagement. *Environment: Science and Policy for Sustainable Development, 51*, 12–23.

Nisbet, M. C., & Newman, T. P. (2015). Framing, the media, and environmental communication. In A. Hansen (Ed.), *Routledge handbook of environment and communication*. Routledge.

O'Neill, S., Williams, H. T., Kurz, T., Wiersma, B., & Boykoff, M. (2015). Dominant frames in legacy and social media coverage of the IPCC Fifth Assessment Report. *Nature Climate Change, 5*, 380–385.

Opitz-Stapleton, S. (2016). *Climate risk and resilience in China*.

Pelling, M. (2010). *Adaptation to climate change: From resilience to transformation*. Routledge.

Saigal, S. (2014). *Policy discourse analysis—India: Supporting climate resilience in policymaking Loodon International Institute for Environment and Developmen*.

Schreurs, M. A. (2010). Multi-level governance and global climate change in East Asia. *Asian Economic Policy Review, 5*, 88–105.

Schreurs, M. A. (2011). Climate change politics in an authoritarian state: The ambivalent case of china. In *The Oxford handbook of climate change and society* (p. 449).

Stockmann, D., & Gallagher, M. E. (2011). Remote control: How the media sustain authoritarian rule in China. *Comparative Political Studies, 44*, 436–467.

Torney, D. (2015). Bilateral climate cooperation: The EU's relations with China and India. *Global Environmental Politics, 15*, 105–122.

Walker, B., Salt, D., & Reid, W. (2006). *Resilience thinking: Sustaining ecosystems and people in a changing world*. Island Press.

Wong-Parodi, G., Fischhoff, B., & Strauss, B. (2015). Resilience vs. adaptation: Framing and action. *Climate Risk Management, 10*, 1–7.

Wübbeke, J. (2011). The power of advice: Experts in Chinese climate change politics. *Fridtjof Nansen Institute (FNI) Report, 15*, 60.

Yi-Jun, C., Ling-Yan, S., & Jieqiong, Z. (2011). National interest and international news reporting: How differently do China daily and the New York Times report the 2009 Copenhagen climate change conference? *Asia-Pacific Science & Culture Journal*, 1.

Zhao, Y. (2004). The state, the market, and media control in China. In P. M. Thomas, & Z. Nain (Eds.), *Who owns the media?: Global trends and local resistance*. Zed Books.

CHAPTER 6

Conclusion

> Mountains, rivers, forests as well as farmlands, lakes, grasslands and deserts all make indivisible parts of the ecosystem. Protecting the ecosystem requires more than a simplistic, palliative approach.
>
> —President Xi Jinping delivered a speech at the Leaders Summit on Climate at 22 April 2021

The above quotation was cited from the presidential speech with a title 'For Man and Nature: Building a Community of Life Together'. The title demonstrates China's latest efforts on developing theories of environmental and climate governance. This signifies that China started to call for building a community of life for mankind and nature. In this sense, the community should be shared by human society and other elements of natural systems. This inherently reflects the key concepts of ecological civilisation which has been prioritised on China's top agenda. Also, the idea for building the community has been constructed on the basis of the concept of a human community for a shared future. The concept is designed to have a better future and a peaceful environment in order to be resilient to the changing international relations. Thus, China has raised

S. Wang, *Climate Change Discourse in China*,
https://doi.org/10.1007/978-981-16-6754-1_6

the concept of ecological civilisation and the call for building the human community for a shared future to achieve the concept of a community of life for mankind and nature.

China's progress in climate governance is on the basis of its national development and ecological discourses. This book has an inherent value of employing a discursive approach to observe how the climate governance has been evolving in China. The various ideas and interpretations of addressing climate issues have been dominant, marginal, adopted and/or even filtered in climate policies. The discursive evolution has been observed across the topics on climate leadership and governing systems, China's role in global climate politics, climate justice and responsibilities and climate resilience.

Summary of Each Chapter

Chapter 2 employs a leadership approach to understanding the transformed climate governance in China. This research offers an innovative perspective for interpreting the leadership and breaks the traditional dichotomy between leaders and followers. Instead, it uses a discursive approach to observing how the climate leadership has been redistributed and shared across different actors. The centralised leadership remains around the top level in China due to the administrative system. However, the leadership has been decentralised and shared across sub-national actors to represent various local voices. Different functional governmental agencies have various focuses and priorities over climate governance. The internationalisation of climate leadership has raised the various actors particularly non-state ones in the context of global climate politics. The shared and redistributed climate leadership has enhanced, rather than undermining, China's low carbon actions.

Chapter 3 focuses on the rise of China in global climate governance. The domestic political discourse on the rise of China in international cooperation makes a significant contribution to Chinese attitudes towards global climate change governance. The role of China in global climate governance had been slightly evolved from 2009 to 2015. In 2015, China's position on participating in international cooperation had outlined the importance of diplomatic relations with major countries, neighbouring countries and developing countries. Accordingly, China had made an effort on working with developed countries, emerging economies and developing countries vulnerable to climate change in terms of international climate negotiations.

Chapter 4 outlines China's fundamental positions on climate change particularly emphasising the principle of common but differentiated responsibilities. This has been institutionalised in the roots of China's climate governance. However, various social actors have been engaged in constructing climate justice and responsibilities. China has been discursively constructed as a major economy and carbon emitter and thus it has been required to undertake the responsibilities for substantial climate actions. This discursive change has challenged the fundamental principle of climate governance of China.

Chapter 5 employs the analytical approach of framing to understand how climate resilience had been constructed in newspapers in China. Compared to mitigation, climate adaptation and resilience have received week attention from a wide range of social actors. This work shows a potential to have discursive evolution of climate resilience in China. An interesting finding is that climate resilience had been interpreted and framed differently between the Chinese-written and the English-written newspapers. Also, while the term 'resilience' had been discussed in the context of urban construction, it has not yet explicitly linked to extreme weather events and natural disasters. In this sense, there remains a discursive gap between climate resilience and climate governance.

Contributions to Academic Theories

This book contributes to the theoretical development of discourses applied in the case of climate governance of China. Hajer (1995) defines the discourse as a set of ideas and concepts which can be constructed and transformed in the political realities. Litfin (1995) employs the discursive approach to observe the framing of science and the environment. Climate change discourses can be socially constructed by various actors (Bulkeley, 2014). Following the existing literature and theories, this book outlines the discursive elements of China's rise in global climate governance and the principle of common but differentiated responsibilities. Various actors have been identified in the discursive evolution of climate governance of China.

Also, this book, particularly Chapter 5, employs the framing approach to have empirical findings of climate communication in newspapers in

China. The framing approach has been applied to studies on climate politics and communication (Nisbet & Newman, 2015; O'Neill et al., 2015; Trumbo, 1996). This work employed the concept of frame to define and identify the problems, causes, moral judgements and remedies of issues (Entman, 1993). An important contribution to the framing approach is recognising the different interpretations of climate resilience in different contexts and between different languages.

Environmental politics can be understood through identifying and summarising different discourses such as ecological modernisation and sustainable development (Dryzek, 2005). This book shows a distinctive feature of climate politics and discourses in China. For example, the discourses 'development' and 'responsibility' are categorised in Chapter 4 and they reflect the context of China in climate governance. This breaks the categorisation of environmental politics developed in the Western countries.

An agent-centred approach has been demonstrated across the book to identify the various actors in climate discourse and governance of China. While the governmental agencies play a key role in the policy-making of climate change, various actors including business actors, international organisations and NGOs have contributed to the development of climate discourses. This approach supports a perspective of interpreting the climate governance of China beyond the governmental policies.

Contributions to Climate Governance

Aligning the governing system with climate objectives secures a sustainable climate action. The case of China demonstrates the importance of adjusting and improving the governmental administrative systems in terms of climate governance. The transferring of the climate leadership in functional agencies from the CMA, the NDRC to the MEE shows a progress in China's reform in climate governance. China has witnessed its rise in global climate governance and has recognised its key role in international climate negotiations. While China's climate actions require a long-term transformation of economic, industrial and energy structures, it insists on the fundamental positions on climate politics particularly the principle of common but different responsibilities.

Climate justice and responsibilities for climate actions have been evolving in the global governance over decades. On the one hand, China is a major economy with a rapid economic growth and a high level of carbon emissions per year. It is very promising to see that China has declared the targets of the 2030 carbon peak and the 2060 carbon neutrality. China has been more proactive to undertake the responsibilities in international affairs including climate change issues. On the other hand, it is a fact that China is fundamentally different to the developed countries in terms of national circumstances and historical responsibilities for climate issues. The developed countries have responsibilities for and capabilities of providing financial and technological assistances to developing countries. For example, the developed countries promised to raise $100 billion in climate finance for adaptation by 2020. However, it remains unclear about how the promise would be translated into actions. China has been engaged in assisting other developing countries to fight against climate disasters. But, in a sharp contrast to the responsibilities of developed countries, China's climate actions on working with other developing countries have been defined as South-South cooperation.

A distinctive contribution from this book is recognising climate resilience in the social context of China. While Chapter 5 focuses on the frames of climate resilience in newspapers and has implications for studies on environmental communication, it shows a gap between the concepts of resilience and the actions of addressing climate issues. The terms 'resilience' and 'adaptations' have different implications for climate policy design and implementation. However, they have not yet been distinct clearly from each other in the Chinese context. A lack of a clear and consistent definition of climate resilience fails to help mobilise a strong support for governance of climate resilience.

Future Development of Climate Governance

Ecological civilisation has been emerging as a key theme across different sectors in China. Even in 2020, while an unprecedented COVID-19 crisis started to storm across China and the world, President Xi Jinping had emphasised the importance of green recovery in economic development. Particularly, the 2030 carbon peak and the 2060 carbon neutrality pave the way for sustainable development and green recovery.

The concept of building a community of life for mankind and nature makes a great contribution to theoretical developments of and practical solutions to international environmental and climate governance. President Xi had attended the General Debate of the 75th Session of the United Nations General Assembly, the UN Biodiversity Summit and the

Climate Ambition Summit in 2020. Also, being invited by the US, President Xi attended the Leaders Summit on Climate in April 2021. China shows a full support for global climate cooperation.

However, challenges for achieving a successful climate action remain around this planet. China's climate objectives require profound industrial transformation and energy transition. Thus, it is very important to continue to observe how climate discourses will be evolving across the topics on leadership, justice, international cooperation and resilience. For example, the term 'resilience' has been rushing into China's public and political discourses. Anyway, it is very exciting to see that China is translating its climate discourses into actions. And, we need to ask: What is the next?

References

Bulkeley, H. (2014). Revisiting... Discourse coalitions and the Australian climate change policy network. *Environment and Planning C: Government and Policy, 32,* 957–962.

Dryzek, J. S. (2005). *The politics of the earth: Environmental discourses.* Oxford University Press.

Entman, R. M. (1993). Framing: Toward clarification of a fractured paradigm. *Journal of communication, 43,* 51–58.

Hajer, M. A. (1995). *The politics of environmental discourse: Ecological modernization and the policy process.* Oxford: Clarendon Press.

Litfin, K. T. (1995). Framing science: Precautionary discourse and the ozone treaties. *Millennium: Journal of International Studies, 24,* 251–277.

Nisbet, M. C., & Newman, T. P. (2015). Framing, the media, and environmental communication. In A. Hansen (Ed.), *Routledge handbook of environment and communication.* Routledge.

O'Neill, S., Williams, H. T., Kurz, T., Wiersma, B., & Boykoff, M. (2015). Dominant frames in legacy and social media coverage of the IPCC Fifth Assessment Report. *Nature Climate Change, 5,* 380–385.

Trumbo, C. (1996). Constructing climate change: Claims and frames in US news coverage of an environmental issue. *Public Understanding of Science, 5,* 269–283.

Appendix A

Table A.1 Framing climate resilience in *People Daily*

Month/Year	*Frames*			*Sources*	*Events*
	Sectoral	*Remedial*	*Moral*		
June 2018	Cities	–	–	–	–
February 2016	Cities	Infrastructure	–	NDRC; MoHURD	Climate policy
April 2014	Regional; Industrial sectors	Development; Disaster management;	Vulnerability	CMA	IPCC report
November 2013	–	Adaptation	–	Switzerland	Climate policy
May 2013	Agriculture; Water	Disaster management	–	Hunan Meteorological Agency	–
November 2012	Industrial sectors	–	–	Shaan Xi Province	–
May 2012	–	Disaster management	–	CMA	–
April 2012	–	Development; Disaster management	–	CMA	IPCC report

(continued)

S. Wang, *Climate Change Discourse in China*,
https://doi.org/10.1007/978-981-16-6754-1

Table A.1 (continued)

Month/Year	*Frames*			*Sources*	*Events*
	Sectoral	*Remedial*	*Moral*		
November 2011	Agriculture; Ecosystem	Development; Disaster management; Infrastructure; Security	Vulnerability	CMA	National policy
November 2011	Agriculture; Costal	Infrastructure	–	MoST, CMA and CAS	Climate report
April 2011	–	Adaptation	–	Tsinghua	National policy
December 2010	Agriculture; Regional	Disaster management	–	CPPCC	Climate conference
December 2010	–	–	Developing countries	–	Climate conference
March 2010	Ecosystem; Forest	–	–	SFA	–
March 2010	Agriculture	Infrastructure; Security	–	CMA	–
September 2009	–	–	Developing countries; Vulnerability	President Hu	Climate conference
September 2009	–	–	Developing countries; Vulnerability	CASS	Climate conference
August 2009	–	Adaptation	–	NPC	International conference
August 2009	Agriculture; Costal; Ecosystem; Forest; Regional; Water	Development	–	State Council NDRC	Climate policy
June 2009	–	Adaptation	Developing countries	NDRC	Climate conference
November 2008	–	–	Developing countries	Premier Wen	Climate conference
July 2008	Agriculture; Coastal; Ecosystem; Public health; Water	Disaster management; Infrastructure; Security;	Vulnerability	CMA	International conference

(continued)

Table A.1 (continued)

Month/Year	*Frames*			*Sources*	*Events*
	Sectoral	*Remedial*	*Moral*		
July 2008	–	Adaptation	–	People's Daily	–
June 2008	Agriculture; Water;	Disaster management	–	–	Climate policy
December 2007	–	Adaptation	–	UNDP	Climate conference
December 2007	–	–	Developing countries; Vulnerability	Sierra Leone; UNFCCC	Climate conference
November 2007	–	–	Developing countries	Premier Wen	International conference
September 2007	–	Adaptation	Developing countries	President Hu	International conference
April 2007	Agriculture; Public health; Water	Disaster management; Security	–	CMA	IPCC report
March 2007	Coastal	Adaptation	–	MoST; People University	Premier press conference
April 2006	–	–	Vulnerability	CASS	International conference

Appendix B

Table A.2 Framing Climate Resilience in *China Meteorological News*

Month/Year	*Frames*			*Sources*	*Events*
	Sectoral	*Remedial*	*Moral*		
January 2018	Ecosystem	Disaster management	–	NDRC	Environmental policy
August 2017	–	Disaster management	–	MoST	–
April 2017	Cities; Ecosystem; Infrastructure; Water	Disaster management	–	Baicheng County, Xinjiang	–
	Cities; Ecosystem; Infrastructure; Tourism; Water	Disaster management	–	Yueyang, Hunan Province	–
	Cities; Forest; Infrastructure	Disaster management	–	Bazhou, Xinjiang	–
	Cities; Ecosystem; Infrastructure	Development; Disaster management	–	Anyang, Henan Province	–

(continued)

S. Wang, *Climate Change Discourse in China*,
https://doi.org/10.1007/978-981-16-6754-1

Table A.2 (continued)

Month/Year	*Frames*			*Sources*	*Events*
	Sectoral	*Remedial*	*Moral*		
March 2017	Cities; Infrastructure	Disaster management	–	Shihezi, Xinjiang	–
	Agriculture; Cities; Ecosystem; Infrastructure	Disaster management	–	Chaoyang, Liaoning Province	–
February 2017	Regional	Disaster management	–	CAAS	–
August 2014	Agriculture; Ecosystem; Tourism	Disaster management	–	CAS	Climate report
	–	Disaster management	–	Kunming, Yunnan Province	
August 2014	Infrastructure	Disaster management	Vulnerability	CMA	Climate policy
	Agriculture; Community	–	–	CAAS	
May 2014	–	Disaster management	–	CMA	IPCC
January 2014	Agriculture; Water	Disaster management;	–	Guizhou, Guiyang Province	–
November 2007	–	Disaster management	–	CMA	Climate report
September 2007	Agriculture	–	–	Ningxia	Climate policy
June 2007	Regional	Disaster management	–	MoST	Climate policy
May 2007	–	–	Developing countries	Oxfam	Climate report
April 2007	–	–	Vulnerability	CMA	IPCC report
April 2007	–	Adaptation	–	CMA	IPCC report
March 2007	–	Adaptation	–	CMA	–

Appendix C

Table A.3 Framing Climate Resilience in *China Business News*

Year	*Frames*			*Sources*	*Events*
	Sectoral	*Remedial*	*Moral*		
December 2015	–	–	Developing countries	UNEP	Climate conference
December 2015	–	Disaster management	–	EU	–
November 2015	Agriculture	–	Developing countries	–	Climate conference
	–	–	Developing countries	NDRC	
October 2010	–	–	Vulnerability	DRC Congo	Climate conference
September 2010	–	–	Developing countries	MoFA	Climate conference
December 2009	–	–	Developing countries	Brazil; Oxfam	Climate conference
June 2009	–	–	Developing countries	UK; Climate Group	Climate conference
December 2007	–	–	Developing countries Vulnerability	Oxfam; Uganda	Climate conference
December 2007	–	–	Developing countries	Oxfam; India	Climate conference
November 2007	–	–	Developing countries	UNDP	International report

S. Wang, *Climate Change Discourse in China*,
https://doi.org/10.1007/978-981-16-6754-1

Appendix D

Table A.4 Framing Climate Resilience in *China Daily*

Date	*Frame*			*Sources*	*Events*
	Sectoral	*Remedial*	*Moral*		
November 2018	–	Development	–	NCN	Climate conference
November 2018	Coastal	–	–	(PEMSEA) IGO	–
October 2018	–	Development	–	MoC Commerce	–
August 2018	–	–	Vulnerability	Difference Group	–
June 2018	–	Disaster management	–	UIBE	–
April 2018	–	Security	–	UB	–
March 2018	Ecosystem; Regional	–	–	TU	–
March 2018	–	Infrastructure	–	WMO	–
	–	Development	–	CMA	–
March 2018	Water	–	–	IR (NGO)	International event
March 2018	Agriculture	–	–	Kenya	National policy

(continued)

S. Wang, *Climate Change Discourse in China*,
https://doi.org/10.1007/978-981-16-6754-1

Table A.4 (continued)

Date	*Frame*			*Sources*	*Events*
	Sectoral	*Remedial*	*Moral*		
February 2018	–	Infrastructure	–	C40 climate change alliance	International conference
February 2018	–	Development; Infrastructure	–	Santa Clara (a local city, Silicon Valley, US)	International conference
January 2018	Agriculture	Development	–	UN	International conference
January 2018	Public health	–	Vulnerability	CUHK	–
December 2014	–	–	Developing countries	UNFCCC	Climate conference
December 2014	–	–	Developing countries	ODI	International report
December 2014	–	–	Developing countries Vulnerability	Australia	Climate conference
December 2014	–	–	Developing countries	China Daily	Climate conference
December 2014	–	Security	–	EU	–
December 2014	–	–	Developing countries Vulnerability	UNDP; UNEP	Climate conference
November 2014	–	–	Developing countries	NDRC	Climate conference
November 2014	–	–	Developing countries	China Daily	International conference
October 2014	Public health	Disaster management	-	IFRC	International conference
September 2014	–	–	Developing countries	UNFCCC; MCEE (NGO); University of Kassel	Climate conference
September 2014	–	–	Developing countries	China Daily	Climate conference
September 2014	–	Disaster management	–	UNFCCC; CCC	Climate conference
September 2014	Coastal	–	–	China Daily	–
September 2014	–	Disaster management; Infrastructure	–	IFPRI	–

(continued)

Table A.4 (continued)

Date	*Frame*			*Sources*	*Events*
	Sectoral	*Remedial*	*Moral*		
August 2014	–	Development	–	UN	–
April 2014	Community	–	Developing countries Vulnerability	Oxfam	IPCC report
April 2014	–	Security	–	IPCC	IPCC report
March 2014	–	Disaster management	–	US	Climate report
February 2014	–	–	Vulnerability	–	–
December 2007	Ecosystem	Disaster management	–	HSBC	Climate report

Appendix E

Abbreviations

C40 Climate change alliance	a network of global cities addressing climate change issues
CAAS	Chinese Academy of Agricultural Sciences
CAS	Chinese Academy of Sciences
CASS	Chinese Academy of Social Sciences
CMA	China Meteorological Administration
CPPCC	The Chinese People's Political Consultative Conference
CUHK	Chinese University of Hong Kong
Difference Group	a non-governmental organisation
IFPRI	International Food Policy Research Institute
IFRC	The International Federation of Red Cross and Red Crescent Societies
IR	International Rivers (NGO)
MCEE	Molina Center on Energy and Environment (NGO)
MoC	Ministry of Commerce
MoFA	Ministry of Foreign Affairs
MoHURD	The Ministry of Housing and Urban-Rural Development
MoST	Ministry of Science and Technology

S. Wang, *Climate Change Discourse in China*,
https://doi.org/10.1007/978-981-16-6754-1

NDRC	National Development and Reform Commission
NPC	The National People's Congress
ODI	The Overseas Development Institute
PEMSEA	Partnerships in Environmental Management for the Seas of East Asia
SFA	The State Forestry Administration
Tsinghua	Tsinghua University
TU	Tibet University
UB	The University of Bristol
UIBE	The University of International Business and Economics
UN	The United Nations
UNFCCC	The United Nations Framework Convention on Climate Change
UNDP	The United Nations Development Programme
UNEP	The United Nations Environment Programme
WMO	World Meteorological Organization

Index

S. Wang, *Climate Change Discourse in China*,
https://doi.org/10.1007/978-981-16-6754-1

The manufacturer's authorised representative in the EU is Springer Nature Customer Service Centre GmbH, Europaplatz 3, 69115 Heidelberg, Germany. If you have any concerns regarding our products, please contact ProductSafety@springernature.com

Printed and bound by CPI Group (UK) Ltd, Croydon, CR0 4YY

29/04/2026

02099478-0016